# MORAL ACQUAINTANCES AND MORAL DECISIONS

T0238975

# Philosophy and Medicine

## VOLUME 103

*Founding Co-Editor*
Stuart F. Spicker

### Senior Editor

H. Tristram Engelhardt, Jr., *Department of Philosophy, Rice University, and Baylor College of Medicine, Houston, Texas*

### Associate Editor

Lisa M. Rasmussen, *Department of Philosophy, University of North Carolina at Charlotte, Charlotte, North Carolina*

### Editorial Board

George J. Agich, *Department of Philosophy, Bowling Green State University, Bowling Green, Ohio*
Nicholas Capaldi, *College of Business Administration, Loyola University, New Orleans, New Orleans, Louisiana*
Edmund Erde, *University of Medicine and Dentistry of New Jersey, Stratford, New Jersey*
Christopher Tollefsen, *Department of Philosophy, University of South Carolina, Columbia, South Carolina*
Kevin Wm. Wildes, S.J., *President Loyola University, New Orleans, New Orleans, Louisiana*

For further volumes:
http://www.springer.com/series/6414

# MORAL ACQUAINTANCES AND MORAL DECISIONS

## RESOLVING MORAL CONFLICTS IN MEDICAL ETHICS

*by*
STEPHEN S. HANSON
*University of Louisville, KY, USA*

 Springer

Stephen S. Hanson
Department of Philosophy
Louisville, Kentucky 40292
USA
stephen.hanson@louisville.edu

ISBN 978-90-481-2507-4 (hardcover)          e-ISBN 978-90-481-2508-1
ISBN 978-94-007-0522-7 (softcover)
DOI 10.1007/978-90-481-2508-1
Springer Dordrecht Heidelberg London New York

Library of Congress Control Number: 2009927031

© Springer Science+Business Media B.V. 2009, First softcover printing 2011
No part of this work may be reproduced, stored in a retrieval system, or transmitted in any form or by
any means, electronic, mechanical, photocopying, microfilming, recording or otherwise, without written
permission from the Publisher, with the exception of any material supplied specifically for the purpose
of being entered and executed on a computer system, for exclusive use by the purchaser of the work.

Printed on acid-free paper

Springer is part of Springer Science+Business Media (www.springer.com)

*This book is dedicated to all those who helped me along the way, in large and small ways. Special thanks are due to H. Tristram Engelhardt, Jr., whose continual support for this project helped lead to its completion. Finally, I give my love to Caroline, Camille, and Lena.*

# Acknowledgement

I would like to thank the original publishers of the following works for their permission to use portions of these works in this book:

Portions of chapter two were published in an earlier form as "Engelhardt and Children: The Failure of Libertarian Bioethics in Pediatric Interactions." *Kennedy Institute of Ethics Journal* 15 (2) June 2005: 179-198. Copyright The Johns Hopkins University Press 2005.

Portions of chapter five were published in an earlier form as "Moral Acquaintances: Loewy, Wildes, and Beyond." *HEC Forum* 19(3) 2007: 207-225. Copyright Springer 2007.

Portions of chapter five come from a paper entitled "Richard Zaner and 'Standard' Medical Ethics," written for the festschrift for Richard Zaner, edited by Osbourne Wiggins, to appear in the same series as this volume.

# Contents

# Chapter 1
# Justifying Moral Claims in a Pluralistic Society

Moral disagreement is to be expected in modern, pluralistic societies. Rational and reasonable persons can and do disagree about the appropriate solutions to various moral problems. Though each party to a moral disagreement may derive moral answers to moral problems in a morally justifiable fashion, different persons will not necessarily arrive at similar moral conclusions. This results from the combination of the under-determination of morality by reason and the related fact of rational moral pluralism. So, when persons who disagree encounter one another in a situation in which they must determine the right thing to do, they may not agree upon the appropriate solution, even when all parties actually desire to reach a solution. In a pluralistic society, where such persons can and do encounter each other in cases of moral conflict, there is a need for, but a lack of, a means of resolving these conflicts in a way that can be morally justified to all participants. Any serious attempt to address moral methodology in a pluralistic context must take seriously the fact of rational moral disagreement.

Though the lack of moral agreement is not a novel concept, the amount and significance of the disagreement is often not fully appreciated. This work seeks to show that it is in fact not possible to resolve many moral conflicts between persons who have deep disagreements, and that moral debaters will need at least narrow areas of agreement for productive solutions to moral conflicts. It will then ask what sort of agreement is possible, and what sort is required, in given cases of disagreement. That reason underdetermines morality is assumed as a starting position; the goal of this work is to explore what follows from that presupposition.

The organization of the argument herein is as follows: in this chapter the project is laid out, a number of key concepts are defined, and the central questions, as well as some questions that will not be asked, are described. In Chapters 2, 3, and 4, I consider as means of addressing the central concern of this work, the resolution of moral disagreement in the context of rational moral disagreement, three theories of medical ethics: H. Tristram Engelhardt, Jr.'s, "content-free" proceduralism through the principle of permission, the four-principle approach of Tom L. Beauchamp and James F. Childress, and the casuistry defended by Albert Jonsen, Stephen Toulmin, and others. Each of these methods seeks to provide tools for persons in a secular pluralistic society to engage in moral debate and even to solve moral problems through discussion with persons with whom they disagree. I argue that, as they are

S.S. Hanson, *Moral Acquaintances and Moral Decisions*, Philosophy and Medicine 103,
DOI 10.1007/978-90-481-2508-1_1, © Springer Science+Business Media B.V. 2009

presented, none of these methods can both give adequate tools for persons who disagree to determine what action one ought to perform in most morally problematic situations and give a justification for that action that persons who disagree ought to think is a plausible justification. This is perhaps a demanding definition of "resolution of moral problems," but it is what is required to satisfactorily address and resolve a moral problem in a pluralistic society: without the first part of this resolution, one cannot know what to do, and without the second, persons involved would not have good reasons to think it was the right thing to do. (Unless otherwise noted, this is what is meant by "resolution" of moral problems in this work.) None of the tools provided by these theories are adequate to perform both tasks in the context of pluralism. I suspect, though I do not attempt to prove, that no theory can do both of these tasks in a pluralistic context. However, I argue that the latter two types of theory can do so in the context of a "moral acquaintanceship," a concept that is discussed in depth in Chapter 5.

The work of H. Tristram Engelhardt, Jr., since his publication of *The Foundations of Bioethics*[1] in 1986, has done much to push forward the debate about moral decisions in a secular pluralistic environment. His arguments, more strongly than most if not all others, challenge any attempts to reduce moral disagreement to mere factual disagreement or to potential agreement based on shared general concepts or principles. Hence it is with Engelhardt's work that I shall begin, as his arguments for the virtual certainty of irreducible moral disagreement provide much of the motivation for this work.

In Chapter 2, I argue that Engelhardt's procedural method cannot successfully conclude that even the limited actions it seeks to justify are the right actions.[2] I further argue that Engelhardt cannot justify the principle of permission and his minimal principle of beneficence to the satisfaction of the very high standards that he sets for justification of moral claims in a secular pluralistic society. I show that they fail to adequately describe, and thus to adequately resolve, difficult cases in bioethics; I also argue that the grounds on which he justifies these principles—as the only ones possible for a minimal moral grammar in a secular pluralistic society—better defend only a different principle which I call the principle of reason-giving, that unfortunately is even less effective than Engelhardt's principle of permission at giving resolutions to moral problems. Consequently, I argue, Engelhardt has posed a problem that his own theory cannot resolve.

I then approach principle-based theory and casuistry as two different means of attempting to resolve the problem of secular pluralism that Engelhardt has

---

[1]Engelhardt, H. Tristram, Jr. (1986). *The Foundations of Bioethics*, 1st ed. New York: Oxford University Press.

[2]While it is true that Engelhardt's theory does allow persons to recognize that the limits of justifiable political power mean that they must allow others to perform some actions, even though their own moral views may show an action to be wrong, this is not all that Engelhardt argues. His theory argues for a secular account of morality, by which we may conclude that certain actions are morally wrong, at least insofar as moral authority can bind one in a secular and pluralistic context.

identified.[3] By an appeal to a specified interpretation of the four principles of medical ethics, or by a careful analysis of paradigm cases, each of these systems provides methodologies that are intended to be adequate for persons to argue that a certain action is morally correct and justifiable in a given case. Principles fail, however, to be shared in specific enough forms to address difficult moral issues in a pluralistic environment. General principles are widely shared, but are not specific enough to assist in resolving difficult moral problems.

Casuistry attempts to resolve problems through an appeal to paradigm cases on which there is broadly shared agreement as to the correct decision. By comparing a current case by analogy to those cases, persons are supposed to be able to derive a solution to a particular case and justify that decision to others who might initially disagree based upon the agreement on those paradigm cases. However, I show that casuistry must overestimate the agreement on paradigmatic cases as well as the potential for arguments by analogy to be broadly shared in order to function as a means of obtaining agreement on a case resolution in a pluralistic society.

Given the failure of each of these theories to resolve the problem posed by Engelhardt's strong interpretation of the multiplicity of moral views in a secular, pluralistic society, I re-evaluate what can be obtained from each theory. In Chapter 5, I argue that, while Engelhardt's theory remains significantly damaged by his own critical analysis, one can use the concept of moral acquaintanceships, a concept derived from Engelhardt's dichotomy between moral friends and moral strangers, to enable practical use of principles and/or casuistic analysis in certain settings.[4] Such "acquaintanceships" can be found in human nature, particular cases, or in public policy, and where they exist, they can allow a resolution of moral problems by providing both the tools to give an answer to the problem at hand and the means to justify that solution as the correct one to all in the acquaintanceship. I conclude by exploring the nature of agreements within such acquaintanceships.

This brief overview of the project is of course limited. A more careful description is provided in *Three Attempts to Resolve Moral Problems*, below, after a refinement of the scope of this work and the definition of important terms.

---

[3]It is not necessarily the case that this is precisely how the authors of these theories would state the goals of their theories. I believe that it is a fair interpretation of their goals; but even if it is not, this is the way in which these theories are often used. It is still relevant that they are not capable of accomplishing this particular goal. In Chapter 5 it is argued that they can achieve a similar goal in more limited settings.

[4]See Loewy, Erich. (1997). *Moral Strangers, Moral Acquaintance, and Moral Friends*. New York: State University of New York Press, and Wildes, Kevin Wm., S.J. (2000). *Moral Acquaintances: Methodology in Bioethics*. Notre Dame, IN: University of Notre Dame Press. Their views, which differ importantly, are discussed in depth in Chapter 5. Though Engelhardtians might be willing to agree with some or all of the conclusions derived by an appeal to moral acquaintances, I believe it does represent a departure from Engelhardtian theory. This is also discussed in Chapter 5.

## Moral Disagreement in a Secular, Pluralistic Society

This is a work in moral philosophy, but it is targeted at a specific area of applied ethics. Medical ethics, perhaps more than many other philosophical fields, has seriously attempted to grapple with the problem of pluralism in ethics. This is likely due in part to the political nature of many bioethical matters, but it also owes much to the work of foundational authors in modern bioethics. Most modern theories of bioethics, from the principles of the Belmont Report forward, are best understood as attempts to address moral matters in a secular, pluralistic society. Arguably, part of the reason for this is that these were the sorts of cases that physicians, theologians, and early ethicists in a medical context were encountering—patients, physicians, family members and/or government officials with multiple importantly contrasting views of what is right. One may consider as examples the cases of Karen Ann Quinlan[5] and Dax Cowart,[6] as well as, e.g., *Roe v. Wade* or *Griswold v. Connecticut*.

Yet in these cases a resolution was derived, albeit not to the satisfaction of all. Further, the Belmont Report, though it certainly acknowledges the existence of differences in moral views, implies that there is a significant commonality that may override these differences. This implies optimism towards the difficulties of pluralism that may not be justifiable. The problem of resolving moral problems in the context of multiple moral views may be even greater than these cases suggest—the commonality implied by the ability to resolve court cases or derive ostensibly shared principles may be more than can truly be presumed. Court cases, especially, do not necessarily imply commonality. Often, they are decided by an appeal to a recognized authority, and not moral agreement—*Roe v. Wade* is an obvious example of such a case.

This extent of this difficulty can be shown through observing a recent case that differs from many others like it only by being quite well publicized. The case of Terri Schiavo reflects a significant divergence of values. Perhaps few in the United States do not know of the case of Ms. Schiavo, who on February 25, 1990, suffered a cardiac arrest, apparently caused by a potassium imbalance, leading to brain damage due to lack of oxygen. On February 11, 2000, it was ruled that she would have chosen to have her feeding tube removed, and after prolonged proceedings involving all three branches of government at both state and federal levels, her feeding tube was permanently removed, and she finally died on March 31, 2005. There were many times when the facts of the case were disputed, but the medical facts of the case were not the issue of importance for many concerned about this case. Many of the public and political disputes over whether it was appropriate to withdraw her artificial nutrition and hydration were not disputes over the facts of the case, but rather disputes over whether her life was intrinsically valuable. Many who held that

---

[5] Supreme Court of New Jersey, "In the Matter of Karen Quinlan, An Alleged Incompetent", in *Ethical Issues in Death and Dying* (1977). 1st ed., Robert F. Weir, ed. New York: Columbia University Press, pp. 274–277.

[6] "A Demand to Die" (1998), in *Cases in Bioethics: Selections from the Hastings Center Report*, 3rd ed. Bette-Jane Crigger, ed. New York: St. Martin's Press, pp. 110–111.

it would be permissible to remove her feeding tube if that was what she would have wanted held such a position because she would have chosen it, and that she would have believed that her life in this state was of no value to her. This is what her husband argued and the courts eventually upheld. Many opponents of withdrawing her feeding tube held that her life was intrinsically important, and that it was valuable whether or not it was currently valuable to her.[7]

The difference is neither a difference in interpretation of facts nor a differing interpretation of shared values, but rather a fundamentally differing conception of what gives meaning and value to a human life. Those who hold one view argue that this value is inherent in human life, and cannot be removed by choice nor destroyed by any non-fatal incident. Those holding one opposing view might argue that individuals can determine for themselves what gives their life value, and that different persons may interpret comparable situations as having differing value. A third view, applicable to this case although not generally argued with regard to it, could be that human life is valuable only if aware, and consequently Ms. Schiavo's life was not of the same value as an aware human's. Other interpretations of the case are also possible. These differing conceptions of the moral value of human life lead to differing appropriate or acceptable solutions to Ms. Schiavo's case, and are not likely to be amenable to a resolution through rational debate.

This brief discussion does not, of course, prove that reason cannot resolve the questions of whether or when feeding tubes may be withdrawn; *a fortiori* it does not prove that reason cannot resolve all or most cases of moral conflict. But it does suggest that it will be at best extremely difficult for anyone to show through reason alone that all persons can be required by reason to agree upon a single answer to the above questions about the right stance to take on withdrawal of life-sustaining therapy, or to similar difficult questions about other issues of moral conflict. If radically different views of the world, including the features that give value to a life, can be held through views that are comparable in consistency and coherence, then even well-exercised reason will be hard pressed to be able to provide a resolution that gels well with each of these varying worldviews, much less to find one answer to which all persons are rationally mandated to agree. Moreover, this comparison shows why moral resolution will be so difficult—the differences between persons can and do run very deep.

The point here is not to *argue* that shared concepts of reason cannot resolve moral conflict. That is a different project. Rather, the argument in this work will show that even if one operates under the assumption that reason alone cannot resolve such

---

[7]Information about this case derived from: Perry JE, Churchill LR, Kirshner HS. (2005). "The Terri Schiavo Case: Legal, Ethical, and Medical Perspectives." *Annals of Internal Medicine* 143(10):744–748; Wolfson, J. (2005). "Erring on the Side of Theresa Schiavo: Reflections of the Special Guardian ad Litem." *The Hastings Center Report*, 35(3):16–19; Koch, T. (2005). "The Challenge of Terri Schiavo: Lessons for Bioethics." *Journal of Medical Ethics* 31(7): 376–378; Dresser, Rebecca. (2005). "Schiavo's Legacy: The Need for an Objective Standard." *The Hastings Center Report* 35(3):20–22; "Schiavo Case Tests Priorities of GOP" Shailagh Murray and Mike Allen. The Washington Post; March 26, 2005; A.01.

conflicts, there still can be a way for persons in a modern, liberal, pluralistic society to resolve these troubling conflicts, even in the face of differing comparably rational views about core issues.

## Definitions

Several central concepts used throughout this work need definition or clarification. The phrases "shared concepts of reason" or "reason alone" shall mean the shared understanding of reason that includes the rules of logic (e.g., identity, *modus ponens*, and so forth) as well as requirements of consistency and non-self-contradiction.[8] No content is involved in these shared concepts; they form solely a set of procedures and rules. This conceptual set should also include a requirement that claims have adequate support to be justified, and some very basic understandings about what sort of claims constitute reasons that form satisfactory justifications. This is not intended to describe the full extent of reason, but rather is meant to be an uncontroversial description of what persons share in their understandings of reason.

The term, "reasonable," as used, for example, in the phrase "multiple reasonable moral views," is meant as a limiting term, in that one is not required to take into account all possible moral views, but only those that are reasonable. However, this limitation is understood in this work in a very weak and structural (as opposed to content-oriented) sense, as too strong a definition of what is reasonable would allow one, intentionally or incidentally, to ignore moral views that one finds inconvenient. As such, the definition of "reasonable moral view" should be taken here to mean a moral view that is composed of claims that can be shown to be recognizably moral, and that follows basic rules of the shared concepts of reason discussed above. Some elements of basic coherence will follow from this, as multiple explicit contradictions or internal inconsistencies will not pass the test of rules of elementary logic, and so it may also be understood that a reasonable moral view is at least minimally internally coherent.

An example of an unreasonable moral view on this account could be a system that holds that "All humans are created equal and endowed with the right to life, liberty, and the pursuit of happiness," while at the same time holding that (non-voluntary) human slavery is morally acceptable. These two concepts simply cannot be consistently held together, as slavery violates the previously stated right to liberty. However, a system that holds that "All *white* humans are created equal and endowed with the right to life, liberty, and the pursuit of happiness," that allows slavery of non-white humans, is not unreasonable on its face, though defending this very ad hoc addition will almost certainly be impossible to do without eventual incoherence.

---

[8] Or, at least, no recognized self-contradiction, given that it is likely that many large sets of beliefs contain conflicting claims that have not been considered contemporaneously, and so the contradiction has gone unnoticed. Reason, as understood herein, will require that such contradictions, once noted, be addressed in a fashion adequate to resolve the contradiction.

The phrase "the problem of moral pluralism" shall be understood to mean the phenomenon whereby, though one may rationally know/understand that a particular action X is immoral or unjustified, one is unable to prove that fact to another person who holds different, reasonable, moral beliefs. For example, Mary K. may hold that it is morally acceptable to test cosmetic products on rabbits for the purpose of creating a new lip liner, and she may be able to produce a consistent moral explanation for why that is so. Nevertheless, just because a consistent moral defense of this claim is provided, if Henry S. holds a different consistent and reasonable moral view that does not allow the moral testing of cosmetics on sentient animals, Mary K. will not be able to convince him that she is right by employing her set of arguments and reasons. Likewise, the reasons that Henry S. can produce to show why one ought not perform such tests on bunnies may not serve to prove that claim to Mary K. Though he could argue that rabbits are sentient beings and will suffer pain from the testing involved, if Mary K.'s coherent system of beliefs consistently holds that neither of these facts is morally relevant then his reasons do not prove her wrong in terms which she must recognize as correct. For example, she might hold that moral rights and moral status come from one's capacity to engage in the practice of granting others rights in a society to remain out of a Hobbesian state of nature. In such a case, she understands that the sentience of a being is not morally relevant, and rabbits need not be granted moral status of any sort. There is no disagreement on the facts of the matter, but only on the moral relevance of those facts. Henry S. could attempt to argue her out of her position, but if that failed he would have no recourse by which to prove her wrong according to her understanding of what makes something wrong.

The problem of moral pluralism is an implication of reason's underdetermination of morality in a modern, pluralistic environment. With regard to topics where persons reasonably believe that an action is morally right or wrong, others can and will reasonably believe otherwise. Each person is logically coherent; they simply hold different views about what sorts of things are morally relevant, or how those moral features interact. As long as those views meet basic minimal criteria for being rational views—e.g., the shared concepts of reason—reason does not determine what those views must be; therefore, one cannot prove either party to be irrational for holding their particular views.

The term "moral situation" should be understood to mean a situation where reasonable persons can recognize potentially morally relevant features that can be affected in ways either good or bad by some persons' actions. Moral situations are often complicated in a pluralistic society by different persons privy to a given situation recognizing different features of the same situation as morally relevant, and/or disagreeing as to the relative importance of different moral features or the perception of given features as moral. These situations will be referred to as "morally problematic situations" or "morally difficult situations," even if, in the end, there is a satisfactory way to address the situation that avoids serious problems or disagreement. It is in these morally problematic situations that reasonable persons who hold rational, but different, basic moral beliefs may disagree about what actions are correct to take in a given moral situation, and may therefore disagree about whether particular actions are justified in such morally difficult situations.

Finally, by "resolving morally problematic situations" or "resolving moral problems," is meant presenting and defending a particular action as the right one to take in a potentially morally difficult circumstance, and justifying that decision as correct to others in a way that those others can recognize as a valid justification. This is an intentionally formulaic construction, meant to be empty of particular content.

## The Problems of Concern in This Work

The ultimate goal of this work is to show that some theories, including but not necessarily limited to principles and/or casuistry, can be used to resolve medical moral problems in the context of a moral acquaintanceship. These acquaintanceships can be found within broadly morally pluralistic societies like modern western societies. Medical cases are not the only sorts of morally problematic cases that one may encounter in a pluralistic society, nor are particular case judgments the only sorts of issues one will need to address in such a society. But case judgments in medical issues are a main focus of the well-known attempts to address these sorts of question discussed herein, and with good reason. Modern medicine is a part of most of the most important features of modern life—birth, injury, health, pain and its avoidance, death. Persons dealing with morally difficult cases of birth, injury, or death are generally more concerned with coming to a satisfactory resolution of their particular situation than they are about formulating rules with broad application. Questions about public policy and other, broader questions are not less important, but they are often not the issues of greatest concern to persons involved in a morally problematic case. To the persons involved in the case, generally speaking, resolving the morally problematic situation at hand is primary.

The sorts of cases targeted herein are the sorts of cases encountered commonly enough in modern medicine that they need to be addressed to practice medicine in a pluralistic environment. These sorts of cases tend to involve persons with differing religious views or ethical commitments, who have differing views on what approach to a particular medical problem ought to be taken. These problems include, but are not limited to, questions of whether one may impose treatment of various sorts on persons who refuse those treatments for various ethical reasons, and whether the society which imposes those treatments can give valid arguments for preferring the rejected treatment, whether the ethical values of parents and guardians may be used to determine treatment for juveniles and other persons not competent to make their own medical decisions, and whether and how the ethical values of the health care team and the society at large—insofar as such values can be determined—can or should impact a particular case judgment.

## Three Attempts to Resolve Moral Problems

It is now possible to give a more in-depth description of the structure of the book. It begins with analysis of the secular theory of H. Tristram Engelhardt, Jr., who distinguishes between interactions between moral friends and moral strangers. "Moral strangers" are:

> . . .persons who do not share sufficient moral premises or rules of evidence and inference to resolve moral controversies by sound rational argument, or who do not have a common commitment to individuals or institutions in authority to resolve moral controversies.[9]

By the terms herein, they hold reasonable moral views, share concepts of reason as defined above, but nonetheless disagree upon enough moral claims or concepts of reason to render them unable to resolve a given moral problem or set of moral problems. Moral strangers are thus defined both herein and by Engelhardt in relational and relative terms, as persons can be moral strangers with some persons but not with others, and moral strangers with one another at one point in time but not at others. The potentially vast difference between persons who hold different rational moral beliefs means that moral strangers do not share a thick, contentful moral view.[10] Different understandings of what sorts of things have value, and of how that value is appropriately appreciated, can make moral views incompatible at their most basic levels. This results in the need for two different types of approaches to moral problems. "Moral friends,"[11] who share a common set of moral beliefs, can share a contentful moral worldview that could serve as a comprehensive moral guide, but they do not share this view with "moral strangers." Between moral friends, the shared moral content in the shared worldview serves as the basis for making and justifying moral decisions; between moral strangers, there is no shared moral worldview, and so one cannot justifiably use one's contentful worldview to justify solutions to moral problems. One needs, instead, an ethics useful for moral strangers to resolve morally difficult cases. He argues that in such cases there is a "content-free" procedural morality that can be employed by all moral persons, because this basic procedural morality is a necessary precondition for moral conversation in the context of moral decision-making with moral strangers. That is, if one is interested in the moral resolution of controversies in a pluralistic environment—and though one is not rationally required to be so interested, most of us are—then, when attempting

---

[9] H. Tristram Engelhardt, Jr. (1996). *The Foundations of Bioethics*, 2nd ed. New York: Oxford University Press, p. 7.

[10] The term "contentful" means "having a significant amount of moral content," similar to Engelhardt's usage of the term "content-full." See, e.g., H. Tristram Engelhardt, Jr. (1996). *The Foundations of Bioethics*, 2nd ed. New York: Oxford University Press, p. 7, where he contrasts content-full morality with procedural morality. I avoid that term here, as it may be misinterpreted to imply that the moral view being described must be "full" or "complete," which can suggest that all possible moral issues have been addressed by the view. A contentful moral view may not address all possible moral issues, but must address a significant amount.

[11] Defined as, "those who share enough of a content-full morality so that they can resolve moral controversies by sound moral argument or by an appeal to a jointly recognized moral authority whose jurisdiction they acknowledge as derived from a source other than common agreement." Ibid. The latter portion of the definition is best understood as a specification of the former, as the source for accepting the legitimacy of such a recognized moral authority can only be an appeal to their shared content-full morality. This is shown in Chapter 5, esp. in *The Separation of the Moral and the Social*.

to resolve a moral problem with others who are moral strangers, regardless of one's contentful views, one can employ Engelhardt's "content-free" procedural morality and its "principle of permission." Further, if one is interested in resolving moral conflicts in a pluralistic setting without recourse to force, then one must appeal only to these content-free procedural principles.

Moral strangers are therefore able to resolve at least some—if not all—of the moral conflicts that are likely to occur between them by appeal to this procedural morality. Not all may find the solutions very palatable, but they can still be understood as resolved in the only way possible. Even when the use of the principle of permission to address moral conflict results in what one takes to be a morally repugnant conclusion—say, for some, the discovery that one cannot justify any prohibition or restriction of abortions in a secular society[12]—moral persons must abide by those conclusions when dealing with moral strangers. Moral friends can understand and justify the moral repugnance of such a conclusion by appeal to their shared moral worldview, and when dealing with other moral friends they can appropriately call it immoral; but if the morality for moral strangers provided by the content-free procedures cannot show it to be immoral, then between consenting moral strangers it cannot be prohibited.

The resulting theory is a very libertarian one. Persons should be left free in a secular pluralistic society to pursue their lives as they see fit, restricted only by the principle of permission's exhortations to "Do not do to others that which they would not have done unto them, and do for them that which one has contracted to do." This is sometimes shortened to, "Do not use other persons without their permission."[13] Within the guiderails of those limitations, persons may be guided by their own contentful moral views, or lack thereof, in their interactions with moral strangers, but have no further obligations to them.

In Chapter 2, I show that Engelhardt's account fails on two counts. First, his own arguments that the principle of permission is a necessary condition for morality do not meet the criteria he sets out for the basic conditions of moral justification, and neither of the arguments that Engelhardt has put forth to prove that it is a necessary moral principle succeed. The arguments in Chapter 2 show that the principle of permission is a substantial moral claim, and that it makes moral claims that cannot be required of rational moral persons interacting as moral strangers. Second, the principle of permission that Engelhardt offers as the only possible means of resolving moral conflict with any appeal to a common moral authority is, in fact, often incapable of resolving moral conflict. If my arguments are correct, the principle of permission is neither required, nor is it very helpful in resolving moral problems.

Though I argue that Engelhardt's project fails, it does so in a way that suggests where one may look to attempt to complete the project. There is no content-free

---

[12]H. Tristram Engelhardt, Jr. (1996). *The Foundations of Bioethics*, 2nd ed. New York: Oxford University Press, pp. 253–258.

[13]Ibid., pp. 67 and 123. Engelhardt also enjoins moral strangers to abide by a principle of beneficence, but this principle is so limited as to be a non-factor in nearly all debate amongst moral strangers. See Chapter 2, *"Content-Free" Ethics.*

principle to which one can appeal for common moral authority among complete moral strangers; but among "partial moral friends"—for example, those that, following Wildes and Loewy, I call "moral acquaintances"—there may be significant agreement that can entail contentful principles. Such agreement, if it can be discovered, will make reason much more effective in resolving moral conflict. Instrumental reason cannot make headway without some beginning claims to work with. But if persons in a pluralistic society can be shown to share some basic, beginning moral beliefs, then those beliefs can potentially be used to create a means for resolving moral conflicts to which all such persons will agree. If that means is created only with instrumental reason and shared basic beliefs, then all persons who share those basic beliefs should agree to the use of that means. It may be concluded that, following Engelhardt's argument, there can be no significant contentful agreed-upon beliefs shared across all rational persons; but an examination of the reasons why will again lead the project in the right direction to show what can be done.

There are two initial primary sources in the literature to which one might look to discover shared basic beliefs. The first is the "common morality" at the level of principles, to which Beauchamp and Childress look to derive their principles of bioethics. The second is the theory of casuistry, which employs cases themselves as the locus for case resolution, resolving current cases by analogy to previously resolved, similar cases.

Beauchamp and Childress argue that the principles of beneficence, respect for autonomy, non-maleficence, and justice are found in the "common morality," and so are shared by all morally serious persons.[14] Consequently, by appeal to these shared principles, they argue, even Engelhardtian moral strangers should be able to find a common area of agreement that may allow them to resolve potential moral conflicts. Since a concern for all of these basic moral notions is present in the "common morality," and should therefore be shared by morally serious persons, whether they are labeled in Engelhardtian terms as moral strangers or moral friends, there is a significant area of agreement that allows resolution of conflicts. The areas regarding which moral strangers differ can be extreme indeed; but, they argue, that which they share is not insignificant either.

Unfortunately, this broad agreement on principles can be sustained only at a level too general for consistent use in resolving specific problems. Though all rational moral persons may well agree that justice is important, they do not agree on precisely how justice should be defined, much less on how to derive specific consequences from it; though they may all agree that beneficence is important, they do not agree on how important it is relative to, say, the importance of respecting the autonomy of one who is refusing a beneficent act. Moreover, though all may agree on the relevant moral terms, it is not clear that different persons actually share the same meaning for each of these terms. Beneficence can easily mean either, "Do

---

[14]This important phrase, as employed by Beauchamp and Childress, is not defined here and not easy to clearly define. See Chapter 3, especially *Two Versions of the Common Morality,* for significant discussion of this. A "morally serious person" is, roughly, a "reasonable" person (as defined above) who seeks moral resolutions to moral problems.

unto others as you see their good," or, "Do unto others as they see their own good." Though morally serious persons may agree generally upon the concepts embodied by the four principles, they may not agree on the content. Though Beauchamp and Childress may have found a locus for agreement, it does not seem that this agreement is actually going to be adequate to achieve resolutions to moral conflicts that meet the criteria defined in this work.

Specification of these principles, though it may help address issues of purported conflict between the principles and accusations of overgenerality, actually exacerbates the problem. In a case where autonomy and beneficence come into *prima facie* conflict, for example, specification of the two principles can clarify their meaning and resolve the conflict, but in more than one way. Autonomy could be specified as being more important than beneficence in a given case, or just the opposite. One may rationally specify principles to very different results; neither reason nor the process of specification itself requires that the principles be specified in one way rather than another. Therefore, specified principles will not be able to provide an adequate ground for resolution of moral conflict in a truly pluralistic environment. Justification of any moral claim based on these principles may either be impossible (because they are too general to truly justify a claim) or unshared. In smaller settings such principles may be useful in resolving moral problems, if all parties to the conflict agree sufficiently about their meaning, content, and how to derive resolutions from them, but not at the societal level.

One might now suggest that the reason for this difficulty is that the search for agreement was taking place at a level that was still too "high up." Albert Jonsen and Stephen Toulmin and other casuists have argued that, while they frequently note dramatic differences in people's moral beliefs at the level of theories, these differences often do not prevent reasonable persons from arriving at agreed-upon solutions to specific cases.[15] These different persons may be able to find a common area of agreement at the level of cases, and in particular in the kinds of cases that are archetypal cases of a particular action. Based on this agreement on archetypal cases, they argue that agreements in less archetypal cases can be derived by analysis and comparison to these agreed-upon cases.

Though it might seem that the level of cases is precisely a level of serious disagreement in, for example, the Terri Schiavo case, casuists argue that the disagreement at the case level is often less stringent than at levels of higher generality. For example, though arguments in favor of and opposing assisted suicide or euthanasia at the policy level may be strongly opposed, with no hope of resolution, there is very often a lot of agreement (though not universal agreement) on specific cases involving particular individuals, even among supporters of different policy-level positions. Perhaps, the casuist argues, by using the intuitive agreements and insights shared on archetypal cases, one can develop a means to resolve less archetypal and more problematic cases. Even if agreement on basic theories or specific principles could not

---

[15]See, e.g., Jonsen, Albert R. and Stephen Toulmin. (1988). *The Abuse of Casuistry*. Berkeley: University of California Press, pp. 16–20.

be derived from such agreement, this level of agreement would be more than adequate to resolve many moral conflicts that moral strangers encounter at the level of specific cases.

The casuists' approach remains unable to achieve the kind of resolution of moral problems that will be needed to resolve conflicts in a pluralist society, for at least two reasons. While the appeal to archetypes, as modified by relevant differences and compelling reasons to decide other than the archetype, may well suffice to allow a given individual or moral community to resolve moral questions, the amount of individual rational variation introduced to compare and contrast specific individual cases with the archetype cases is sufficient to introduce rational disagreement. Though the agreement on some archetype cases may be near to universal, the understanding of what differences are relevant and what sorts of reasons for deciding a case differently from the archetype are compelling are not universally agreed upon, nor are they definable in ways that all reasonable persons must accept. The agreement on easy specific cases is not enough to provide agreement on difficult ones.

Second, there is a difficulty in determining precisely what cases are archetypal, and to which archetype a given case is most closely analogous. Is a case of physician-assisted suicide most similar to the archetypal case of murder, which is obviously wrong unless there is some very compelling reason to think otherwise, or is it more similar to the archetypal case of a voluntary medical procedure, which is acceptable unless there is some compelling reason to think otherwise? Even if persons can agree upon what sorts of reasons would be sufficiently compelling to think otherwise in either case—which is itself unlikely—a pluralistic group may not be able to determine which archetype ought to be considered. Without these agreements, agreement on the resolution of archetypes cannot suffice for resolution of difficult moral conflicts.

The arguments by analogy that casuists use function in the context of an understanding of what cases are analogous and what features of the comparable cases matter, but they cannot resolve moral conflicts if the analogy itself is not shared. In a society where persons hold different understandings of what is relevant and how relevant those features are, arguments by analogy, including casuistic moral arguments, can only function to allow persons with a fair amount of prior agreement to resolve moral conflicts.

A third problem is that there may not be adequate agreement on the archetypal cases. For while one might think that there is little difficulty getting agreement from all rational moral persons that an archetypal case of murder is *prima facie* wrong, it may be actually much more difficult to figure out what that really means. Is murder the killing of any human? The killing of a born human? The killing of a sapient being, whether human or not? The killing of any sentient being? In order to derive agreement on an archetypal case it is necessary at least to know what the case is to be an archetype of. Often, the agreements on archetype cases that can be derived are themselves based on hidden, but real, assumptions about moral value, which need not be and are not shared across a pluralistic society. These problems may be alleviated by appealing to cases within the context of a more narrow agreement between a smaller number of persons than the whole of a community; but based on these problems, casuistry cannot always resolve problems in a fashion that will work in a pluralistic society, largely because it attempts to find underlying agreements in that society that are not, in fact, going to be there.

## Is There Any Solution?

What is shared in each of these three theories, and in their difficulties, is that they can at best only defend very thin sets of moral views to moral strangers. The broadly shared content of these theories is too thin to resolve effectively the sorts of problems encountered in medical ethics in a pluralistic society. The problem that this seems to display is that the very attempt to justify moral claims to all reasonable persons is doomed to failure. At this point, it is reasonable to ask if there is any way out of this apparent problem.

One could appeal to communitarian thinking, but any appeal to standard communitarianism will prove unable to resolve problems in a pluralistic environment. This is perhaps unsurprising, as moral communities are meant to provide a moral basis within coherent, complex communities with a long history and tradition for the members of that community, not persons in a pluralistic society. However, it may be possible to split the difference between justification to all reasonable persons and justification to a moral community to provide the justification possible within that community while providing the appeal to a pluralistic society which justification to all reasonable persons could provide. Such a solution can be found in the notion of moral friendship and moral acquaintanceship in individual cases, and this hopeful possibility is explored in Chapter 5.

# Chapter 2
# Engelhardt and the Content-Free (?) Principle of Permission

In seeking a means of addressing problems in a modern secular society in such a way that has moral authority for all parties to a given issue, one might search first for a means to eliminate the grounds of conflict. If a theory can appeal to a source of moral authority which all persons must recognize, and that theory can resolve moral conflicts, then such a theory would provide rationally justified means of resolving morally problematic cases. It is a premise of this work that a thick set of moral claims cannot be so justified to all, but it might be possible to justify and defend a minimal set of claims so that at least some reasonable conclusions can be made and justified to all potential parties to a moral conflict. Such a "thin theory" would be a means to derive solutions to moral conflicts with secular moral authority.

H. Tristram Engelhardt, Jr., seeks to derive a very thin, but justifiable in a secular, pluralistic environment, "content-free" moral theory that is capable of resolving moral conflict in a secular pluralistic society. He argues that this theory should be used to resolve moral conflict in medical ethics because it is the only way to do so that can appeal to a secular moral authority that can bind all persons. If successful, he shows that at least some, if not all, morally problematic cases can be resolved by appeal to a content-free and thus universally shared secular moral system. Further, he argues that it is the only such system available for the resolution of moral conflicts in a secular pluralistic society.

This chapter presents a twofold argument. First, I argue that Engelhardt's theory is not well suited for addressing moral problems in medicine, and I show how other, equally well-grounded, theories can be derived that can better address these problems; second, insofar as the system he defends is a moral system, other, possibly more justified, systems can be understood as moral systems as well. Thus, his system will neither be the only means of appealing to a secular moral authority that can bind all persons, nor will it be the most justified or most effective such means for addressing problems of medical ethics.

S.S. Hanson, *Moral Acquaintances and Moral Decisions*, Philosophy and Medicine 103, DOI 10.1007/978-90-481-2508-1_2, © Springer Science+Business Media B.V. 2009

## Engelhardt's Content-Free Theory

Engelhardt addresses the problem of resolving moral problems in a pluralistic society by avoiding the content that, because it is inherently rationally contestable,
engenders moral conflicts in society.[1] The secular moral theory derived through
this method is meant to serve as a means for resolving moral conflicts in a pluralistic society by being the only means of obtaining secular moral authority, because
it contains no content but is based only on pure procedures. It is morality for moral
strangers, persons who do not share sufficient moral premises or rules of evidence
and inference to resolve moral controversies by sound rational argument. These
procedures are, in his words, the basic grammar of morality, the basic concepts necessary to praise and blame persons for their actions in a secular, pluralistic context.[2]
Such a theory could then be used to justifiably resolve moral conflict in a pluralistic
society.

This is only one of Engelhardt's goals. He is equally concerned with showing the
limits of general secular moral arguments[3] and the related limitations of the moral
authority of the state.[4] However, Engelhardt also intends to provide a secular moral
theory of bioethics to which persons can appeal for resolutions of morally difficult
cases between moral strangers. He explicitly states that "*the moral account* justified
in [the *Foundations of Bioethics*] provides a secular moral warrant to authorize coercive force to protect persons when acting peaceably..." and that "what is offered
will still function to secure *a general secular bioethics.*"[5] He is glad to have shown
that "the collapse [of the project of grounding a content-full morality in reason] is
not so complete as to despoil us of any remnant of a general secular morality....
There is still a sparse fabric that we can invoke when we meet in the midst of unbe-
lief [i.e., as moral strangers]."[6] He creates and defends what can be called a "moral
stranger morality," by which I mean a moral system according to which, and by
reference to which, one may justify one's actions to a person who does not share
one's contentful moral beliefs. For those with whom one shares contentful moral
views, one may appeal to those contentful views to resolve moral questions and
to justify those resolutions, but with someone who does not share some or all of

---

[1] In H. Tristram Engelhardt, Jr. (1996). *The Foundations of Bioethics*, 2nd ed. New York: Oxford
University Press. [hereafter *Foundations*, 2nd ed.]

[2] *Foundations*, 2nd ed., p. 70.

[3] *Foundations*, 2nd ed., pp. 8–9, 37–67.

[4] *Foundations*, 2nd ed., pp. 166–179, esp. 177–179, wherein he explains not only the appropriate
role of the state to enforce contracts, limit unconsented-to force, use common resources and provide
mechanisms to resolve disputes, but also gives multiple indications for when the moral authority
of a government is suspect.

[5] *Foundations*, 2nd ed., pp. 11, 12 (emphasis added).

[6] Engelhardt, H. Tristram, Jr. (1997). "The Foundations of Bioethics and Secular Humanism: Why
is there No Canonical Moral Content?" in *Reading Engelhardt: Essays on the Thought of H.
Tristram Engelhardt, Jr.*, Brendan P. Minogue, Gabriel Palmer-Fernandez, and James E. Reagan,
eds. Boston, MA: Kluwer Academic Publishers, p. 280.

those views (e.g., a moral stranger), or someone whose contentful views one does not know (e.g., a potential moral stranger), one cannot appeal to contentful views and expect them to provide justification for oneself and the moral stranger. In such a case, one can appeal only to a "common morality that can bind moral strangers,"[7] a default position which appeals to no contentful views that might be a source of rational disagreement. It is this goal of the *Foundations* that is of the most interest for this work.

This secular morality is limited in its applicability. Engelhardt argues that our contentful moralities are what inform most of our moral decisions. But these various different systems of beliefs are often incompatible with each other, often in very dramatic fashions[8]; moreover, he argues that all attempts to find a contentful moral system that must be held by all persons are unsuccessful, as each of them fails to be a system to which all rational persons must agree.[9]

There is, he argues, still a way to derive a purely procedural morality that avoids all contentful statements, as those can be rationally doubted, yet still can make some clear statements about what sorts of actions can and cannot be morally permitted in a pluralistic society. He argues that a "principle of permission" can be defended on grounds, not of contentful moral claims, but of claims about the bare necessities for the possibility of a secular moral authority.

Engelhardt argues this by process of elimination. He argues that there are four options for resolution of moral conflicts:

> Controversies can be resolved on the basis of (1) force, (2) conversion of one party to the other's viewpoint, (3) sound rational argument, and (4) agreement.[10]

He argues that the fact of moral pluralism shows that conversion will not suffice to resolve all or many moral conflicts in such a society, and any attempt to resolve conflicts by appeal to reason will require an appeal to a particular contentful view, which could not be justified to all members of a pluralistic society.[11] One is thus left with either force or agreement as the only plausible means of resolving moral conflict in a pluralistic society.

Engelhardt also rejects force as a moral means of resolving disputes—though he does allow that some uses of force can be morally justified, which are discussed in *The Rejection of Force* below. He argues that "one must distinguish between resolutions through *cloture* (main force) and resolutions that have moral authority."[12] The argument is not so much that force is ineffective in resolving controversies, but that it is not a moral means:

---

[7]  *Foundations*, 2nd ed., Preface, p. x.

[8]  *Foundations*, 2nd ed., pp. 3–11.

[9]  *Foundations*, 2nd ed., pp. 40–65.

[10]  *Foundations*, 2nd ed., p. 67.

[11]  *Foundations*, 2nd ed., p. 68; see also pp. 35–67 wherein he rejects various attempts to define a contentful universal moral theory.

[12]  *Foundations*, 2nd ed., p. 67, italics in original.

An appeal to force will not answer ethical questions. . .even if the resolution imposes a widespread consensus. Using force. . .will simply be an act of force for any who do not share the moral vision that purports to make such interventions legitimate. . . . To ask a secular ethical question is to seek a ground other than force for resolving moral controversies.[13]

To make a secular moral claim, then, one may not appeal to mere force. This leaves only the final possibility, agreement, by which Engelhardt largely means specific, explicit, contractual agreement. Thus, secular moral authority is granted only through specific consent. It is from this process that the principle of permission that drives the "moral stranger morality" is derived: agreement is the only source of moral authority to which moral strangers can justifiably appeal for moral authority in a secular pluralistic context. When a contentful moral view cannot be derived by reason, and no such moral view is shared, one can still appeal to this default moral position.

From the above argument it can be seen that a "resolution" of a moral conflict is something more than the simple removal of the source of conflict, as brute force could do that. It is also something more than answering a moral question in a way that is acceptable to some, but not all, parties to the conflict. Prior conversion or a particular rational argument could do that. For example, all devoted animal liberationists presumably could agree that xenotransplantation with today's poor results is unacceptable, but not all others would necessarily agree. A resolution of a moral conflict must therefore be a solution to that conflict, an answer to the question "What ought I do (or how ought I act) in this situation?", that does not appeal to unshared contentful moral views, and that has moral authority without the appeal to contentful moral views. Solutions derived from mutually agreed upon claims fulfill this requirement.

Agreement as a source of conflict resolution, however, does not mean only that moral action can occur when there is agreement between all participants. This concept has implications not only for when there is agreement upon what ought to be done, but also for when there is no agreement. The possibility of resolving moral conflict between moral strangers by agreement, in Engelhardt's theory, means a specific principle that outlines moral interactions between moral strangers for when they agree and when they cannot agree: the principle of permission.

## "Content-Free" Ethics

The principle of permission is stated on p. 83 of the *Foundations*: "[O]ne should use persons only with their consent. . .", and more formally on p. 123, "Do not do to others that which they would not have done unto them, and do for them that which one has contracted to do." This is often in tension with beneficence, which presumably could require that one do good for another whether or not one had contracted to do so. Though Engelhardt states that "a commitment to beneficence characterizes

---

[13] *Foundations*, 2nd ed., p. 67.

the undertaking of morality,"[14] only the principle of permission is required for the "very coherence of the moral world," by which Engelhardt means "the possibility of coherent resolution of moral disputes."[15] This entails that secular morality has an "unavoidably libertarian character"[16] in what one can require from others who are moral strangers, and of what is owed to those same others: one may demand of others, and owes to others, only that to which one has contracted, either implicitly or explicitly. Otherwise, unless someone explicitly invites a particular interaction, one is constrained to leave that person alone, unless protecting the innocent from unauthorized force. All other more substantial duties must come from an explicit agreement of one sort or another.[17]

Engelhardt also puts forth a principle of beneficence for interaction between moral strangers, but it is not very substantial. (Between moral friends, on the other hand, the obligation to be beneficent can be and normally is quite extensive.) Any moral obligation of beneficence for moral strangers is essentially limited to what one has contracted to do; though the "principle" of beneficence can be phrased as, "Do unto others their good," it is not incumbent upon one to ever attempt to do to anyone any good at all. If one chooses to, of course, one may; but it is never oblig-atory. This does justify the community as a whole in redistributing any common holdings as welfare grants which may be refused by those who are their targets, but with anything one owns for oneself—one's money, time, body, labor—one has no secular moral obligation, ever, to be beneficent to a moral stranger.[18]

Since one is never obliged to be beneficent by the morality for moral strangers, except insofar as one has contracted to do so,[19] the resulting moral theory thus has no real requirement of beneficence for individual private actions that goes beyond the principle of permission. The role of the principle of beneficence in a secular plu-ralistic society is largely one of public policy implications, as it guides the distribu-tion of any goods held in common by the community. Because one only has respon-sibilities of beneficence to do what one has contracted to do, one has no obligation to do anything for anyone else unless one chooses to. Moreover, the beneficence that it can allow is minimal: though unrequested violence is unacceptable, allowing others to be harmed or to die when one could easily prevent it is not something that can

---

[14] *Foundations*, 2nd ed., p. 123.

[15] *Foundations*, 2nd ed., p. 105.

[16] Engelhardt, H. Tristram, Jr. and Wildes, Kevin Wm. (1994). "The Four Principles of Health Care Ethics and Post-modernity: Why a Libertarian Interpretation is Unavoidable", in *Principles of Health Care Ethics*, Raanan Gillon, ed. New York: John Wiley & Sons, pp. 135–147. See p. 147.

[17] *Foundations*, 2nd ed., p. 123.

[18] *Foundations*, 2nd ed., Chapter 3, esp. pp. 123–124.

[19] The content of a principle of beneficence can come from an implicit contract, wherein one fashions or joins a community with a common view of goods and harms, or through an explicit agreement. In either such case, Engelhardt notes, "the content of a duty of beneficence is grounded in the principle of permission." Consequently, "the principle of permission is conceptually prior to the principle of beneficence." See *Foundations*, 2nd ed., p. 123, under Principle II, Sections A, B, and C.

be prohibited, or criticized, on the grounds of the secular moral theory. Moreover, if one would have to violate other persons' selves or property without their freely given permission, even to do a great good, it is forbidden. There is not even a moral dilemma in this for the theory—it is simply unacceptable to violate the others or their property. This seems to mean that, as James Nelson notes, it would be impermissible "merely to shove an unconsenting, innocent person slightly to the left" in order to avoid some significant tragedy.[20] It also means that one cannot impose even a minimal tax on non-common holdings to provide better health care, fire or police protection, etc., unless explicit permission from all taxees is given.[21] No secular moral condemnation at all can be placed on the miser who hoards food until it rots while people starve outside his gates.

Additionally, one is only restricted by the constraints of avoiding unrequested violence when making one's contracts. One need not accede to any particular rules of conduct when conducting contract negotiations, except that one must avoid acts or threats of violence. One may be manipulative, and is under no obligation to offer reasonable (or any) compensation if one can arrange otherwise. Dramatic differences in power will tend to lead to unequal contracts, but this is permissible as long as both parties agree rationally to the contract, even if there is no realistically viable alternative for one party.[22] This is true if one person, through no desert or effort of her or his own, is a better debater, or more charismatic, or otherwise more powerful. Such differences may be unfortunate for the one made worse off by them, but they are not unfair.[23]

---

[20] Nelson, James Lindemann. (1997). "Everything Includes Itself in Power", in *Reading Engelhardt: Essays on the Thought of H. Tristram Engelhardt , Jr.*, Brendan P. Minogue, Gabriel Palmer-Fernandez, and James E. Reagan, eds. Boston, MA: Kluwer Academic Publishers, pp. 15–29, esp. p. 17. This is, of course, a concern common to any theory that completely prioritizes rights claims over beneficence claims.

[21] See, e.g., *Foundations*, 2nd ed., Chapter 4, esp. pp. 171–175. This stance is ameliorated somewhat by the concept of common holdings, which is Engelhardt's means of reconciling the essentially libertarian theory with the fact of uneven distribution of resources. (See *Foundations*, 2nd ed., Chapter 4, esp. pp. 179–180.) Ideally, one would be able to take of ample resources and create whatever one could, then trade it for whatever one could; in reality, there are no more free resources to take and create with. Since his theory of ownership is based on Locke (see pp. 154–166), some version of the "Lockean Proviso," which holds that when one takes possession of resources, one must leave "enough, and as good, left in common for others," must hold. See p. 158. Since it is no longer possible for a person to leave behind as many resources as she takes, Engelhardt holds that, in order to ameliorate this inability, there ought to be an international tax on land holdings, which could then be distributed for the common good. But even with this qualifier, in the world of Engelhardt's secular moral theory it is impossible to make very many moral claims, as the land tax would be rather minimal, since it is a tax on resources but not on anything done with the resources. For further discussion, see Hanson, Stephen S. (2007). "Libertarianism and Health Care Policy: It's not what you think it is." *Journal of Law, Medicine, and Ethics* 35(3): 486–489.

[22] For a discussion of peaceable manipulation and power differentials, see *Foundations*, 2nd ed., pp. 308–309.

[23] See the discussion of the natural and social lotteries, *Foundations*, 2nd ed., pp. 379–387.

A final point to note about Engelhardt's theory is that contracts and agreements, from which stem all moral obligations in the secular world, can only be made by "persons," who are defined as beings that are "self-conscious, rational, free to choose, and in possession of a sense of moral concern."[24] This means that, unless one contracts with other persons about them, one's secular moral obligations to beings that cannot make contracts for themselves, such as animals, infants, severely mentally retarded adult humans, and the like, are very limited.[25] This has obvious implications for such acts as abortion and infanticide—neither can be prohibited in a secular pluralist society[26]—but also has further troubling implications, which are discussed below in *Case 2.2: Problems with Ownership.*

The moral world described by this theory cannot require beneficence—indeed, it cannot say what beneficence must entail—and, in it, only fully autonomous persons directly matter morally. Many harms and tragic events that might seem unfortunate, and deserving of amelioration, we nonetheless have no moral obligation to rectify, and individuals are not required to assist or deal with others. Abortion, prostitution, gladiatorial combats to the death, animal abuse, persons starving while others gorge in vomitoriums, and so on, all are completely permissible as long as the persons who are participants, for whatever reason, including having no other realistic choices available, have freely chosen to take part.

Engelhardt sees the minimal nature of secular moral authority as a tragic consequence of being unable to defend a contentful theory on secular moral grounds.[27] Yet he argues it is unavoidable in the secular sphere, if persons or societies are to be able to retain the ability to resolve moral conflicts in a way that can justifiably bind moral strangers.[28] He personally believes those activities listed above to be morally repugnant,[29] but recognizes nevertheless that a secular moral argument cannot

---

[24] *Foundations*, 2nd ed., p. 136.

[25] Insofar as one's actions regarding them affect other persons, one can have obligations; for beings such as children and fetuses, actions (such as non-fatal injury) that will impact them when they become future persons can be restricted on grounds of their being unconsented to force against that future person. See *Foundations*, 2nd ed., pp. 143, 146. However, Engelhardt is explicit that one will be "given little satisfaction" in any attempt to assign on general secular grounds some of the rights of persons to human non-persons (infants, the severely mentally retarded, the very senile), even "the right not to be killed nonmalevolently at whim. . . ." *Foundations*, 2nd ed., p. 147.

[26] *Foundations*, 2nd ed., pp. 253–271.

[27] See, e.g., *Foundations*, 2nd ed., p. 13: ". . .it is still possible to save a shred of the modern moral philosophical project, though most (including the author) may find that this salvageable shred is very much less than they had wanted or expected."

[28] *Foundations*, 2nd ed., p. 70.

[29] With the possible exception of animal abuse, on most reasonable definitions — he does, after all, eat veal with full knowledge of what is entailed in the modern production of veal. See *Foundations*, 2nd ed., pp. 144–146, where he appeals to considerations of beneficence to protect animals; but since beneficence is essentially undefined and unrequired outside of a specific moral community (see pp. 123–124), this is at best a minimal protection. See, for example, his mention of hunting and farming on p. 141—hunters and farmers are those who determine how and on what grounds to be "beneficent" to their targets and charges.

condemn them, because only the principle of permission, and all that follows from that, can be justified in a secular moral fashion. In other words, these are cases where one can only say, "X has a [secular] right to do that, but it is wrong," though one cannot prove its wrongness in general secular moral terms.[30] The emptiness of a general secular moral theory is evidence to him that, if one is to understand fully the moral world, know how to treat non-persons, and know how and when to be benef-icent, and so on, one must choose and accept a moral community with a particular contentful moral view. The sparsity of the secular moral ethic means one ought not depend upon it to define the full moral world; nevertheless, that is all that one can depend upon to resolve moral conflicts between moral strangers.

There are two important problems with this analysis, which the remainder of this chapter will show. The first, covered in sections *"Conclusion-Free" Ethics?* to *The Limitations of Contracts*, inclusive, is that, though this theory is premised upon needing a means to resolve moral conflicts between moral strangers, Engelhardt's theory is simply unable to achieve resolution in some of the kinds of cases where resolutions to such conflicts are most needed. This inability to resolve or appro-priately describe difficult moral cases arises frequently in bioethics, because of the nature of the practice of medicine, which makes this lack even more troubling for Engelhardt's attempt to determine the foundations of bioethics.

Still, this result may be acceptable, if disheartening, if he can argue that his theory is the only theory defensible to all and thus the only legitimate option available for resolving bioethical questions in a pluralistic society. Engelhardt does not argue that the principle of permission is a good source of ethics; he argues that it is the only one that can have general secular moral authority. Thus, the above critiques might be accepted as regretfully true. However, this first critique lays the groundwork for a stronger objection by showing some limitations of the theory as it stands and examining the sources of those limitations, which then invites a more fundamental criticism.

The second, more fundamental, critique, developed in the remainder of the chap-ter, is that Engelhardt cannot successfully defend this minimal theory as justifiable to and for moral strangers in a modern, liberal society. The principle of permis-sion is not the minimum necessary principle to engage in moral conversation and allow for a means of praising and blaming. Therefore, it is not the only legitimate option available for resolving bioethical questions in a pluralistic society, but is rather one of several similarly legitimate options, some of which are more content-ful and thus more useful than the principle of permission. This second criticism, in combination with the initial critiques, argues that Engelhardt's theory should be rejected, and a different approach to moral conflict resolution in a pluralistic society sought.

---

[30] See *Foundations*, 2nd ed., p. 84.

# "Conclusion-Free" Ethics? Infants, Ownership, and Unconscious Persons

First, Engelhardt's theory is unable not only to resolve, but also even to address, some important kinds of moral conflicts that arise in modern medical ethics. In each type of case, not only are the tools provided by Engelhardt's secular morality insufficient to determine the correct answer for moral strangers to take, but in fact one cannot take any action, including no action whatsoever, that does not at least run the risk of flouting the principle of permission. Though Engelhardt does propose resolutions to each of these sorts of cases, I argue that these resolutions cannot suffice to resolve these cases to his own standards.

Consider three possible cases:

> **Case 2.1** An adult man is severely injured in a car wreck. When he is found, he is unconscious and in need of resuscitation. Though his breathing is restored, he remains unconscious with an unknown amount of brain damage. He is thought to have a good chance to regain consciousness with careful monitoring and treatment as needed. He has insurance to cover all hospital stays and treatment, but no indication has been made anywhere of his preferences regarding DNR orders or ventilators. No relatives can be found who can help clarify his preferences.
>
> **Case 2.2** An infant falls from its carriage at a shopping mall. He tumbles down some stairs, landing, as luck would have it, directly at the feet of a secular humanist volunteer doctor at a local blood drive. She examines the infant, discovering that he has several severe injuries requiring hospitalization; she begins to stabilize him and calls 911. At this point the parents finish rushing down the stairs and insist that the infant be taken instead to the nearest Christian Science healer.
>
> **Case 2.3** A 14-year-old Jehovah's Witness is brought to a hospital in need of, among other things, a blood transfusion. Since he is not of the age of majority, his parents refuse the transfusion for him; they feel very strongly that this is in his best interests. The doctor feels very strongly that having the transfusion is in the boy's best interests. In fact, he will die soon if he is not given the transfusion. The boy himself is ambivalent. He seems to recognize and respect his religious beliefs, but also explicitly expresses a preference to stay alive.

## Case 2.1: Damned If You Do...

In most normal situations, the principle of permission would allow doctors to provide whatever services the patient wanted and could afford that the doctors were willing to provide. In the first case, though, there is no way to determine what treatment the patient wants. The question is not the philosophical chestnut of whether a sleeping or unconscious person retains personhood, which Engelhardt adequately addresses.[31] Rather, it is the very real question a physician might ask about the patient's preferences for treatment. Doctors cannot know whether he desires no machinery at all to keep him alive, ventilators but no cardiopulmonary resuscitation, full time nursing and immediate resuscitation in all cases, or anywhere in between. They also cannot discover if the possible brain damage would make his choices different (i.e., if he would want resuscitation in most cases, but not if he were likely to be brain-damaged, or how serious the damage would have to be before he would no longer prefer resuscitation.) While various doctors might have their own beliefs on these subjects, they cannot know which is the patient's preference; thus, they cannot know which treatments would violate the patient's free choices and which would support them. They cannot act without danger of infringing on a person's permission; more to the point, as shown below, a libertarian analysis is unhelpful for discovering what one ought to do in this case.[32]

Engelhardt does give instructions on how to treat persons who, like the patient in this case, were once competent to make their own decisions but left no specific instructions as to what decisions to make were they to become incompetent, and who do not have a proxy decision-maker. One is to "choose on [the patient's] behalf either by appeal to a standard of a reasonable and prudent person (within a particular moral community) or by an appeal to a standard articulated by a particular community of individuals committed to a particular set of standards and values."[33] Yet, here there is no indication of the community to which the patient belongs, and thus to which set of community standards or concept of what is reasonable and prudent one should appeal. Absent a specific community understanding of reasonability and prudence, these standards lack content; reasonable and prudent persons could choose very different actions in this case.

This may seem incorrect. After all, to most people, this case seems to be an easy one: when in doubt, resuscitate the patient, and continue life-sustaining treatment unless and until it no longer provides physiological benefit. We do not know the patient's community, but we could act on an assumption that permission would be granted by most "reasonable and prudent" persons, and thus go with the best odds of what action would not violate permission, but this has two problems. First, it isn't clear that such an assumption is valid if there is a danger of significant brain damage, as many people would not prefer to be resuscitated only to be severely brain damaged; and second, the concept of implied permissions and implicit contracts goes beyond the bounds of what Engelhardt can authorize, in part for reasons of avoiding mistaken assumptions. Engelhardt is significantly motivated by the concerns of

---

[31] *Foundations*, 2nd ed., pp. 151–154.

[32] They could act if there were a voluntarily accepted public policy. But that policy would have to be universally accepted, not merely voted into place by majority rule, in order to ensure that acting in accordance with it respects the principle of permission.

[33] *Foundations*, 2nd ed., p. 322.

minority views in society; to retreat to acting on the majority viewpoint here would be inappropriate.

For this reason, Engelhardt specifies who should be considered as a guardian for such a patient, who shall then be able to determine what is "reasonable and prudent" for that patient. "[A]bsent prior agreements, individuals who care for and nurture formerly competent people come to be in authority over them. . ." and consequently have the appropriate authority to choose for them.[34] That guardian will have that authority to choose "because they have a form of property right in their wards as masters have in their indentured servants."[35]

There are two difficulties with this analysis. The first difficulty is with regard to the notion of property rights and ownership of incompetent individuals granting rights of decision-making over them, which are addressed more fully in the following sections *Case 2.2: Problems with Ownership* and *Case 2.3: "Semi-Persons?"* But the second is one that remains problematic even if the former concern can be removed or alleviated. In a discussion of fetuses and infants, who will likely become persons at some point in the future, Engelhardt argues that one can be morally responsible for one's actions against current non-persons (such as non-fatal injury) that will impact them when they become future persons.[36] Such actions can be restricted on grounds of their being unconsented to force—and thus, violations of the principle of permission—against that future person. The same restrictions should apply to actions performed on this patient, who was once a functioning person and probably will be again if treatment is continued and significant brain damage has not occurred.

These considerations lead to an interesting dilemma: depending upon the patient's desires prior to becoming incompetent, continued aggressive treatment might be seen as an application of unconsented to force by the patient once he regains personhood. On the other hand, if one does not treat and the patient never regains consciousness, he will never regain personhood. In such a case, there is no danger of injuring the future person, as the future person will never come to be. But if he were to regain consciousness after a period of non-treatment, having suffered from more severe injury due to non-treatment, then under those circumstances, if the patient had previously had preferences for aggressive treatment, he might also legitimately claim injury due to the actions of the health care professionals. Arguably, this could be claimed not to be an injury caused by action, but rather caused by non-action and consequently not the responsibility of the physicians who did not act, but this would perhaps only push the problem one step back. The patient could not only argue that his preferences for aggressive treatment were not followed, but also that he was harmed by being actively resuscitated but not provided with aggressive treatment, and thus allowed to regain personhood with the injury of not having aggressive treatment.

---

[34] *Foundations*, 2nd ed., p. 322.

[35] *Foundations*, 2nd ed., p. 322. See also pp. 327–328.

[36] *Foundations*, 2nd ed., pp. 146, 258–260.

The only ways to be sure to avoid infringing on a person's permission, and to avoid unwitting violation of permission and injury to a future person, are either to choose, as a health care practitioner, to refuse to deal with such cases, or to choose to avoid resuscitation unless one has a clear indication of the preferences of the patient. Both go directly against the common understanding of what one should do in this kind of case, and are inconsistent with the practice of emergency medicine, and with the law. More importantly, they are not ways of addressing the central difficulty in the case at hand but rather ways of avoiding it. The tools provided by Engelhardt provide no way to both engage in the case and ensure that one avoids the unacceptable action of injury to a future person.

## Case 2.2: Problems with Ownership

The infant is unable to make any sort of choice, and thus is not a person in Engelhardt's sense. It therefore has no general secular moral standing, except as property.[37] It can be treated, as can other non-person things, in whatever way the "owning" persons choose. In a secular society, "non-persons [because they cannot, by definition, be self-legislating] will have imposed on their destiny the particular choices of particular persons or communities of persons."[38] This account may seem troubling already for its implications about the minimal value of infants, but there is a larger problem for Engelhardt: what person or persons should have the right to impose that destiny, e.g., who "owns" the infant? (As noted above, similar problems occur in choosing guardians for incompetent adults.) I argue that, despite providing a clear answer to this question, Engelhardt cannot even employ the rather disconcerting language of ownership to deal with infants, because he cannot answer this question of ownership on secular, pluralistic grounds. Does one allow the infant to be taken to the Christian Science healer, because the parents rightly own the infant, or does one allow the doctor to rush it to the hospital on the grounds that the doctor properly owns it? To assume that either the parents or the doctor, or, for that matter, any random passer-by, have/has the authority to make the responsible autonomous choice for the infant is shown below to presuppose a certain contentful claim or theory of the good. In this case, either (1) it is more important that parents, rather than others, make decisions for their infants than it is to protect the same infants' lives, or (2) protection of life is more important than parental control over decisions for infants. Unless one knows which of (1) or (2) is more important, which one cannot on terms of the principle of permission, one cannot even employ the language of ownership properly in this case. Thus, despite the fact that Engelhardt gives clear rules about the ownership of children, discussed immediately below, he cannot successfully address the moral status of children, even on his own terms.

---

[37] *Foundations*, 2nd ed., p. 165. But see below, where questions about Engelhardt's ambivalence on this topic are raised.

[38] *Foundations*, 2nd ed., p. 141.

Engelhardt does attempt to explain who "owns" children, but his solution cannot be defended in non-contentful terms. He argues that infants, and other incompetent children, are to be treated as the property of their parents because the parents produced them. Engelhardt argues that ownership of objects is granted by placing effort into common property. One converts a "thing" into a "product" by putting labor into it and thus extending one's own self into it. Since one owns oneself, one also owns the products of that self, unless and until one contracts to trade them away.[39] Something becomes one's property when one "enters into a thing, refashions it, remolds it, and... mingles labor with the object" in order to make it an extension of oneself.[40] Engelhardt holds that, since the infant has been produced by the labor of the parents, they own it, and can treat its injuries as they see fit. Thus, like any other thing the autonomous parents have created, the infant is the property of the parents to do with as they will.[41] This puts a human being in the same class as a piece of furniture built in the garage or a meal made in the kitchen, and it seems to allow unacceptable behavior on the part of the "owning" parents. If someone decides to remodel, he or she may smash, sell, burn, or ignore the furniture; if the dinner is bad one may throw it in the disposal and order pizza. Persons may be erratic or wanton in doing these things and be considered no more than eccentric or wasteful, but if children are owned in a similar fashion the results are extraordinary. Parents could beat a child, play football with it, or simply ignore it without doing anything for which another person could hold them morally accountable.[42] As James Nelson has noted, they presumably could even raise it as a veal calf to be fatted for later eating.[43]

Engelhardt does argue significantly that parents may not be malevolent to their children, which might seem to resolve the difficulty until one asks how malevolence is defined. It is defined, as it must be in a content-free account, by the person

---

[39] *Foundations*, 2nd ed., pp. 154–166.

[40] *Foundations*, 2nd ed., p. 157.

[41] *Foundations*, 2nd ed., p. 165. Note that this is a simplified version of the argument, which fails to take fully into account the idea of a "social" sense of personhood, which is discussed in *Social Personhood* below. On pp. 329–330 Engelhardt discusses a very similar case, apparently concluding that the parents can refuse a life-saving blood transfusion for the child, on the grounds that it is not a person and, if not treated, will not become a person. The infant thus is theirs to deal with as they see fit, as long as they do not treat him malevolently—as such is judged only by the parents.

[42] With the not insignificant proviso noted in *Case 2.1: Damned if you Do. . .*: if the child is abused but allowed to live until it grows into a thinking, autonomous person, then that child herself could at that point hold the parents accountable for unacceptable violation of her autonomy by their infliction of physical or psychological injuries that remain now that she has become capable of granting or withholding permission. See *Foundations*, 2nd ed., p. 260: "Choosing for [a child] does not violate the principle of permission, unless it is clear that [her] permission would not be given." But if the parents are sufficiently thorough in their abuse of the infant that they kill her, or if the infant is sufficiently mentally handicapped so that she never develops into a full person, then the parents cannot be held accountable in a secular society.

[43] Nelson, "Everything Includes Itself in Power", p. 17.

who is considering whether an act is malevolent. On Engelhardt's account, this must be the case because there can be no universally accepted contentful secular definition of beneficence or malevolence, and thus the only definition of it possible that could apply to the agent in question is the one that she understands. No particular understanding of what is malevolent to an infant can be understood, with the exception of acts that violate the infant's future potential permission (see note 42). Thus, although there is a fair bit of attention paid to beneficence in the *Foundations*, none of it shows that actions such as those noted above would be impermissible. It is, for example, within a plausible interpretation of malevolence to argue, for example, that infants do not yet have desires for the future and thus are not harmed by a short period of milk feeding followed by (painless?) slaughtering for their meat, or by experimentation on them followed by autopsy to examine the results.[44]

As disturbing as these images may be, the problem for Engelhardt's theory extends beyond the apparent moral repugnance of these actions. One could hold that this repugnance cannot be defended on secular moral grounds and thus hold, as Engelhardt often does, that "A person has a right on secular, pluralistic grounds to do some action 'X,' though I know X to be wrong on the grounds on which I understand morality to depend, but cannot defend in secular terms." The more important problem goes back to the very notion of owning the infant. The problem for Engelhardt is not just that owning a human being in this sense may be disturbing, but more fundamentally that one cannot determine who "owns" the infant without presupposing a particular contentful view. It is crucial to know who "owns" the infant in this case, since that person or persons determine what constitutes beneficent action toward that child. But, I argue, one cannot know whether the doctor or the parents own this infant without presupposing a contentful view. Consequently, Engelhardt cannot employ the language of ownership to infants in secular, pluralist terms; and because he rejects contentful claims in his theory, he cannot employ this language at all.

To begin with, the concept of ownership does not apply to infants very easily. As John Moskop notes,[45] infants and children are unlike any other things that one might own: the labor, and indeed the raw materials, involved in the creation of a child, or a fetus, are quite unlike the labor and raw materials involved in creating most other possessions, making owning children at best an "imperfect fit with other types of property rights [Engelhardt] recognizes."[46] Consequently, figuring out exactly why and how parents are thought to own children takes a bit of work.

---

[44] See the discussion of beneficence in *Foundations*, 2nd ed., on pp. 123–124. Since the infant is not yet a person covered by the principle of permission, there is little to make one think that they can get any protection at all from the prohibition on malevolence except, of course, from persons who understand nonmaleficence to include not harming infants and children. See further discussion below.

[45] See Moskop, John C. (1997). "Persons, Property, or Both? Engelhardt: on the Moral Status of Young Children", in *Reading Engelhardt: Essays on the Thought of H. Tristram Engelhardt, Jr.*, Brendan P. Minogue, Gabriel Palmer-Fernandez, and James E. Reagan, eds. Boston, MA: Kluwer Academic Publishers, pp. 163–174.

[46] Moskop, "Persons, Property, or Both?", p. 171.

Part of the difficulty for Engelhardt's theory is that ownership is not a very apt description of the situation of a pregnant woman and her fetus, or even of a parent and child. Pregnancy involves embodying a being that is partly an other, partly oneself, yet really neither self nor other, growing and developing within oneself in a nurturing relationship rather unlike any other. A pregnant woman is *sui generis*; she is not much like a person who owns an object, nor is pregnancy really very like anything else.[47] Note how attempts to describe pregnancy by comparison to various other relationships inevitably must involve very contorted examples, such as Judith Jarvis Thomson's well-known violinist example.[48] In the same way, questions about whether pregnant women own their fetuses seem to be stretching the concept of ownership to cover something that it really is not much like: ownership of a four-month-old fetus cannot be transferred, for example. The difficulty in asking and answering questions such as those about the ownership rights of women over their fetuses comes at least as much from the fact that pregnancy is not much like any other case of owning something as it does from differing views of what it might be permissible to do to a fetus in a given situation.[49] A different means of describing the relationship is necessary to make sense of the very uniqueness of the situation, and if ownership is the only means available to Engelhardt, then his theory cannot describe these interactions very well.

But even if one were to attempt to describe this relationship in terms of ownership, Engelhardt's theory is not able to show in content-free terms how one ought to understand what it takes to "own" a child. This will cause significant problems for the theory. The relationship between parents and born children is a little more amenable to description as ownership, as the unique embodiment of pregnancy is no longer present. But the ownership description still encounters serious problems.

---

[47] Margaret Little points this out in her forthcoming work: Little, Margaret O. (Forthcoming). *Intimate Duties: Re-Thinking Abortion, the Law, & Morality.* New York: Oxford University Press. This is a claim which Engelhardt disputes—he argues both that this has been the manner of describing the condition of infants and children for much of the history of human existence, and that describing the situation as *sui generis* is a modern invention which requires a contentful theory to function. (Personal conversation, July 2000.) The former claim may be true, but then much of the history of humanity has also allowed adults to be called property as well; it does not follow that either are good descriptions of the relevant beings. The latter claim seems false at least insofar as the circumstances, both social and physical, surrounding the creation of an infant are different from those surrounding the creation of any other thing.

[48] Judith Jarvis Thomson's classic example describes a violinist hooked up to the kidneys of an unknowing and unwilling host. Though it does resemble in some ways an undesired and unsought pregnancy, many readers are immediately struck that this seems to fail to encompass much of what pregnancy is and entails. See Judith Jarvis Thomson's "A Defense of Abortion." Widely reprinted, including in *The Problem of Abortion*, 3rd ed. (1997). Susan Dwyer and Joel Feinberg, eds. Belmont, CA: Wadsworth Press, pp. 75–87.

[49] Thus, for example, a comparison between abortion and the above mentioned cases of maltreatment or killing of infants that Engelhardt could make can be a legitimate one, but it would show only that in both cases ownership is at best a problematic way to describe the relationship between the parent and fetus or born infant. See footnote 47.

It will not do, to begin with, to argue that mere genetic production—e.g., participation in impregnation—entails ownership. One could argue that the germ cells that combined to create the infant were part of the parents' bodies, and thus owned by them, and consequently the infant is their property because it was created from their property. There is some textual evidence to indicate that Engelhardt holds this view: He states in his "Principle of Ownership" that "[y]oung children and human biological organisms are owned by people who produce them."[50] This could be understood to entail that the grounds for this ownership are the production and joining of genetic raw materials. On the Lockean/Hegelian sense of ownership that Engelhardt uses, persons own their bodies more clearly than they own anything else.[51] Anything that one creates from one's property with one's labor is also one's property, and so an infant could be understood to be the property of its parents on these grounds. This would answer the question of ownership of the infant in Case 2.2 in favor of the parents, but this line of reasoning is troubling both in general and in particular for Engelhardt. This view suggests that the effort that goes into gestating and raising the child could not grant ownership, since the infant already is held in joint ownership by the genetic father and mother and, unless explicitly transferred, it remains their property. Thus, a man could impregnate a woman, ignore her for nine months—or nine years!—and retain unchanged joint ownership of the child.[52]

---

[50] *Foundations*, 2nd ed., p. 165.

[51] *Foundations*, 2nd ed., p. 155. It is worth pointing out that Engelhardt's theory of ownership is controversial and that his theory could be challenged on these grounds as well as the moral grounds on which it is often challenged, as Locke's position that property rights derive from labor mixing is commonly criticized. See Wolf, Clark. (1995). "Contemporary Property Rights, Lockean Provisos, and the Interests of Future Generations." *Ethics* 105: 791–818, esp. pp. 794–797. Since I critique Engelhardt's position on its own terms, rather than by challenging his terms, I do not pursue this line of criticism here.

[52] This concern admittedly ignores any issues of "property abandonment," but this concept will not help Engelhardt here. The argument below could be extended to allow the doctor another ground for asserting lack of ownership by the parents, as they arguably have "abandoned" their "property" by refusing treatment.
A second concern with the purely genetic ownership explanation is that it can, in odd cases, imply that a non-person has ownership of an infant. It is not unknown for severely retarded, or even comatose or upper-brain dead women to "somehow" become pregnant; on this account, she would own half the fetus and infant. If, of course, one were to argue that her ownership does not exist because she is not a person, then one would be left with the even less appealing claim that the father has full ownership—but more importantly, it isn't clear how one could claim that. If genetic production of germ cells entails ownership of what is created via that germ cell (assuming said germ cells are not voluntarily given to another) then a non-person seems to have the right to ownership of an infant. The non-person mother did, after all, create the egg cell in exactly the same way as any other pregnant human. It is reasonable to suggest that ownership would lapse if a current person who is a parent were to become upper-brain dead, but where in this case does the lapse occur? Well before the impregnation, perhaps, but how can ownership of a thing lapse before the thing itself exists?
An additional issue along the lines of ownership by non-persons is raised by James Rachels in "Why Animals have a Right to Liberty", Regan, Tom and Peter Singer, eds. (1986), *Animal Rights*

Moreover, although there are some textual grounds to think that this is how children are owned by parents, this interpretation runs contrary to other statements Engelhardt makes on the topic: "One also owns what one produces. One might think here of both animals and young children. Insofar as they are the products of the ingenuity or energies of persons, they can be possessions."[53] This seems to preclude ownership of infants based solely on prior ownership of genetic source material and suggests that active efforts to raise the infant are what grant ownership. This might seem to exclude parental ownership of embryos in some cases, such as prior to knowledge of the embryo's existence; still, this view may be the most consistent with Engelhardt's full understanding of property.

Ownership of infants based solely on prior ownership of one's germ cells seems contrary to the way that Engelhardt interprets his Lockean/Hegelian notion of property, which is that something becomes one's property when one "enters into a thing, refashions it, remolds it, and, to follow Locke's suggestion, mingles labor with the object" in order to make it an extension of oneself.[54] This is a more accurate description of parenting than it is a description of producing and joining germ cells.[55] It would seem reasonable, then, to suggest that, though genetic relationship may imply that one is likely to be the owner of an infant, and perhaps that that relationship gives one *prima facie* standing as an owner, actual ownership must come not from genes but from labor. The question is: what labor grants ownership of an infant?

It could be understood as mere gestation, although I think this would strike most adoptive parents—and, probably, most other parents—as insufficient to constitute parenting. It could be actually raising the child, which is where one can shape and influence a child most. But, at least theoretically, it also could be some other sort of labor, and this possibility points to the crucial question. At what point does one exert sufficient effort to be able to call an infant one's property, and how does one fail to do so? In the case at hand, I argue that both the parents and the doctor can make a plausible claim that they have successfully put sufficient labor into the infant to "own" it, and each can justifiably can claim that the other has failed. If so, then both have a plausible case, and one cannot justifiably decide between the two on secular, moral, non-contentful grounds.

Although initially it seems that ownership of an infant must go to the parents, as Engelhardt clearly believes,[56] such a claim cannot be defended on the grounds he sets. According to Engelhardt, a thing is owned when violating it would, in an important sense, violate the person who owns it.[57] The owner has brought the owned thing into his or her "sphere." This is crucial to the theory, because this is how private

*and Human Obligations*, 2nd ed. Englewood Cliffs, NJ: Prentice Hall, Inc., pp. 122–131, where he argues that Locke's theory of ownership shows that squirrels must own the nuts that they gather.

[53] *Foundations*, 2nd ed., p. 156.

[54] *Foundations*, 2nd ed., p. 157.

[55] See Moskop, "Persons, Property, or Both?", pp. 170–171.

[56] *Foundations*, 2nd ed., pp. 148–149, 156, 327.

[57] *Foundations*, 2nd ed., p. 164.

property rights are extended from the basic injunction of the principle of permission against unconsented-to interference: one acts against the person of the owner when one acts against her property because ownership makes that thing, in an important sense, a part of her. Who, in this case, has brought the child into their "sphere," and who has failed? The answer cannot be derived without an appeal to a contentful notion of what is important, which cannot be defended in general secular terms. This means that, contrary to his claim that the parents are the legitimate owners, Engelhardt cannot explain who owns this infant.

It is true that the parents can claim that they have exerted the necessary effort to own the child in a way that the doctor has not. They have raised the infant, fed it, changed its diapers, and so forth, and begun to exercise their preferred value system upon it. All of this labor has made the infant a part of them, of their sphere, and to remove it from them would be to harm them as persons. The doctor, having just encountered the child for the first time, cannot, of course, lay claim to having done any of these things, except perhaps the last action of exercising a value system on it.

But this last action is important in this case, for the doctor has begun efforts to save the infant's life by the best medical methods available, which, in her value system, one must do in order to care for the child properly. She will have to cease these life-saving efforts in order to relinquish control of the child to the parents. Providing the best medical treatment available when needed is, to her, a crucial part of the labor of raising a child, which the parents are explicitly refusing to do. Had the accident not occurred, she would have had no cause to challenge the ownership rights of the parents; but since it has, and they have refused to permit her to perform this, to her, necessary effort, on her view they have failed to exercise the relevant effort necessary to bring the child into their sphere, and thus cannot own it. By dint of her effort to save the child's life, the doctor may now claim that she has brought the child into her sphere as a patient in need of her help, and because of her dedication to preserving human life, to allow the child to die by stopping this effort would harm her, the doctor, as a person. By placing her effort into the unowned object that is the infant—unowned because the parents have failed to exert the necessary effort to truly own it—she has made it her own.

The above discussion makes clear that neither the parents nor the doctor is clearly wrong in asserting exertion of the appropriate labor to own the infant, and it is not my aim to defend one side or the other. All that is needed is to show that the resolution of who owns the infant cannot be achieved without appeal to some grounds on which to decide what labor is relevantly important to the ownership of an infant; and what labor actually is important may differ from person to person, depending upon their contentful views of what matters in the world. Even if describing fetuses, infants, and other future persons as owned things is not itself problematic, the account of infants as property cannot explain to whom these infants belong, nor who ought to determine correct treatment for them.

This is not to say that societies do not manage to address these kinds of questions; of course, they do. Laws are established that might say—although not in these terms—that the infant "belongs to" the parents and that the "work" put into the infant by the doctor is insufficient to imply ownership. Of course, the laws would not be absolute; presumably, there are cases in which treatments should be performed on children against their parents' wishes and cases in which they should not. Part of the goal of an ethical theory should be to help determine when the imposition of a treatment is appropriate and when it is not. But Engelhardt cannot help with this. He can accept neither the decision to treat nor the decision not to do so because neither can be justified without an appeal to unshared, contentful claims, and this is unacceptable on his account. Engelhardt's theory cannot resolve this problem on its own terms and thus cannot tell one how to act, either in this case or more generally in terms of developing laws to govern such issues. Ownership as an account of relationships between adults and infants cannot provide any means to resolve problems in medicine involving fetuses, infants, or young children.[58]

## Social Personhood

One might note that thus far no reference has been made to Engelhardt's concept of "social personhood," with which he softens the impact of his view and grants to children and infants rights very similar to the rights of actual persons. Social personhood is a means of granting some moral standing to certain non-persons, either as individuals or as groups, because of the particular standing the individual or group has in our society. Infants and young children, for example, are granted nearly all the rights of full human persons in our society. Actual persons grant them these rights not because they are persons, but rather because the practice is useful for preserving certain qualities in our society that are desirable for actual persons. Specifically, society grants humans in these groups a social, "honorary" personhood because doing so

(1)... support[s] important virtues such as sympathy and care for human life, especially when human life is fragile and defenseless. In addition,... (2) [this] offer[s] a protection against the uncertainties as to when exactly humans become persons strictly, as well as protecting persons during various vicissitudes of competence and incompetence, while (3) in addition [it] secur[es] the important practice of child-rearing through which humans develop as persons in the strict sense.[59]

In other words, actual persons grant infants, severely retarded adults, and other human non-persons rights nearly equivalent to those of full persons, in part in order to protect their own interests and in part because it promotes traits they value.

However, the concept of social personhood, although interesting, will not afford an Engelhardtian a response to the above critiques. The practice of social personhood is justified as a protection for actual persons. "Since this [social] sense of a person cannot be justified in terms of the basic grammar of morality (i.e., because such entities do not have intrinsic moral standing through being moral agents)..." it must

---

[58] Engelhardt does defend a principle of intervention to protect children and other wards, to which he could appeal. This principle will not help in this case. See discussion in *Engelhardt's Principle of Intervention* below.

[59] *Foundations*, 2nd ed., pp. 147–148.

be justified through these consequentialist grounds.[60] Therefore, social personhood could not countermand the requirements of the principle of permission in a given case. Violating the principle of permission involves imposing unwanted force on an unwilling other, which cannot be justified on the grounds of desired results. Furthermore, as shown below, the only way to defend social personhood on purely secular moral terms is insufficient to include infants, and thus social personhood could not apply in this case, even if the concept were adequate to oppose the parents' permission.

If the rights of a social person directly come up against the principle of permission—e.g., if the preceding consequentialist grounds require using someone's property, and by extension using that person, without his/her permission—it is impermissible to respect the granted rights of social persons at the expense of actual persons' rights. Freedom is a side constraint in Engelhardt's theory, and cannot be trumped for any purpose.[61] So, for example, when Moskop argues that Engelhardt's notion of owning children is balanced against the rights of those children, granted by social personhood, and notes that this is not so far off from what one normally thinks is reasonable for how children should be treated, he is mistaken to think that the two have comparable justificatory value.[62] Consequently, although Engelhardt's grounds for social personhood are good reasons to treat all humans as persons, they cannot justify interference with an actual person, which is precisely what either the parents or doctor could argue is occurring in the case at hand. Because of their superior foundation, even the least of a person's liberty-based, side-constraint property rights always will supersede any social personhood rights of children. Because of the way in which Engelhardt views and justifies these rights, social personhood never will amount to much more than saying, "We actual persons choose to treat you non-persons in this fashion; but we could always change our minds with no penalty or wrongdoing."

Although Engelhardt attempts to strengthen the concept of social personhood by holding that much can be done by threats of social isolation if one violates the mores of one's society, this response misses the point. If parents do not mind social isolation, or can perform their abuse without being caught, then society can do nothing further to hold them responsible for anything they do to their infant property. Social personhood, justified, as it must be, without direct reference to the principle of permission, always will be subordinate to that principle, and thus the most minimal rights of actual persons always will trump the strongest socially granted rights of non-persons.

An additional question arises even if one seeks to grant the rights of social personhood to the infant in this case. Which of the infant's conflicting socially based rights, the right to life or the right to religious freedom, ought to be respected? This question probably could be resolved in favor of life in this case because, although

---

[60] *Foundations*, 2nd ed., p. 147.

[61] *Foundations*, 2nd ed., p. 70.

[62] Moskop, "Persons, Property, or Both?" pp. 171–172.

a child may be a member of a religion by birth, baptism, and the like, she cannot be a self-chosen member of that religion. Because of the importance of individual choice in a libertarian theory, choice of religion is probably important for religious freedom in Engelhardt's theory, but some effort still would be needed to parse out precisely which of a social person's rights should be acted upon.

Finally, and most importantly, social personhood for infants cannot be required by the principle of permission. Of Engelhardt's three justifications for social personhood, only the second is directly related to the principle of permission. When it is not clear whether a being is actually a person, but one knows that it soon will be, a careful approach, such as the acceptance of social personhood, is arguably reasonable. But the first and third justifications are plausible only on particular contentful views of the good. Protecting human lives and developing persons are both admirable goals on many theories, but neither is directly related to the principle of permission, which only applies to actual persons. If an actual person had no desire for sympathy or care and/or had no desire for future humans to exist, one might pity him, fear him, or think he was hideously mistaken; but one could not, on secular moral grounds, condemn him as morally wrong. Engelhardt must hold that, although this person may be wrong according to one's particular moral views, he has the right to be so wrong.

Thus, if social personhood can be defended at all on secular moral terms, it is only in terms of the second justification. And, whatever else is true, it is clear that infants and toddlers do not have what Engelhardt requires for personhood. Before acquiring language at the level required by Engelhardt for personhood, which most infants do not come close to for at least a year, they cannot possibly qualify as persons in the strict sense, and thus the uncertainty about when a human becomes a person is nonexistent in non-lingual humans, at least. The second justification thus does not extend the concept of social personhood to all humans, including the infant in the current case. Consequently, Engelhardt's concept of social personhood, like the concept of ownership of persons, provides no resolution to Case 2.2.

## Case 2.3: "Semi-Persons"?

Case 2.3 involves a semi-autonomous "semi-person," one who is able to think and make certain decisions, but who presumably is unable to understand completely and decide properly in this case, and hence is unable to give permission for treatment or non-treatment. The law, which treats the unemancipated 14-year-old as unable to make his own health-care decisions, makes this presumption; the parents and the doctor also make it when they each attempt to make decisions for the patient. The situation is more complicated than that of the infant, since a 14-year-old is much more able to understand the situation, and yet presumably is less able to do so than an adult.

Engelhardt does provide a means to resolve cases involving adolescents, but his method proves unsatisfactory. He argues that parents may act as guardians, in particular with regard to selecting or rejecting health care for their children, based on either the fact that they produced the child (the ownership relationship described above) or the "indentured service" nature of the relationship between caregiver and care-receiver.[63] In other words, the parents choose what care the adolescent receives, either because the parents own their child, and thus can do with him as they will (which seems inappropriate, since the child exhibits some of the important qualities of strict personhood), or because their relationship with him implies that he must submit to their will in matters of his care. Engelhardt's argument is that, insofar as an older child has developed the abilities of personhood, he implicitly has chosen to exchange obedience for the support given by a family. Insofar as he has not acquired the abilities of personhood, he remains property. In either case, the child is to obey and submit to the will of the parents.

Though this may seem initially appropriate because the adolescent is not a full person, this seems to describe the situation poorly. Although such an adolescent is a person, or nearly so, in many respects, he is not yet a fully autonomous one.[64] It seems that the principle of permission ought to apply in some fashion to this "nearly person," since the alternative of placing all decision-making authority for this human entirely in someone else's hands seems inappropriate, given that the child has, by definition, some capacity for making decisions. But he is not a full-fledged person and so cannot take the full Engelhardtian responsibility for making the decision in this case. Any appeal to ownership by another person is made even less plausible than in the first case by the adolescent's partially autonomous character. In short, there seems to be no one able to grant permission in this case, as none of the players—neither the patient nor the parents nor the health-care providers—have sufficient right and ability to make an autonomous decision for the 14-year-old. If true, then no decision can be justified on Engelhardt's own secular pluralistic grounds.

Engelhardt's trichotomy of options in this case—children are either owned by their parents, indentured servants to them, or fully emancipated—is insufficient to appropriately describe these kinds of interactions between adults and children. The most applicable category is that of "indentured service," but this is a troublesome category. To begin with, it is not clear what basis there is for believing that parent-child relationships do develop, or should develop, in such a way that the child is in a period of "indenture" to the parents, particularly to the extent Engelhardt takes it. Engelhardt suggests that the caregiver/care-receiver relationship entails that the child must accept his parents' wishes in his care, but there is little argument to support this conclusion. Even if one accepts this interpretation of the relationship, however, Engelhardt cannot argue successfully that such a relationship entitles the parents to refuse lifesaving therapy. It is true that there is a general understanding in society that, "While you're living under my roof, you'll follow my rules," including

---

[63] *Foundations*, 2nd ed., pp. 155–156.

[64] Were he fully autonomous in all respects it would be appropriate in that case to treat him as a full person, fully emancipated from his parents. Since such a case would be fairly straightforward for Engelhardt's theory, this case is defined in such a way that the adolescent is explicitly not able to be fully emancipated.

things like curfews, rules on dating and sexual interaction, and rules on the use of various mood-altering substances. Furthermore, there is some latitude for individual variation in different houses, but such house rules would never be understood generally to include something like, "As long as you are under my roof, you will refuse life-saving treatment, and perhaps die, if I think that is the best thing for you, even if you do not." Some specific argument would have to be made, but is not, to show that this strong of a rule could be justified in any such "indenture."

In part, this is because it does not seem to be an implicit part of any theory of indenture that the "master" has the choice of life or death over the "servant." In slavery, e.g., complete ownership, this may be so, but not in a normal indenture. Compare the other examples of indenture given by Engelhardt to parallel this relationship: joining the Marines or the French foreign legion, entering a marriage or monastery, or interning at a hospital.[65] Granted, joining military groups explicitly does involve risking one's life in combat, but that possibility explicitly is (or should be) made clear. Based on Engelhardt's own examples, it is unlikely such a possibility is implicit in the other cases of indenture. The contract that a hospital intern accepts does not require her to treat patients in such a way that she knowingly will be exposed to a lethal dose of radiation, especially if another alternative, such as wearing protective clothing, is available; nor may an abbot force a monk to starve himself.

It is true that interns and other physicians are required to treat persons even if exposure to those persons might risk their health. Surgeons are obligated to perform operations on patients infected with HIV or Hepatitis C, thus risking accidental blood-to-blood contact. Physicians (and other health care workers) of all sorts are obligated to be on the medical front lines in the case of epidemics, whether they be relatively known risks such as bird flu, or complete unknowns such as the initial outbreaks of Legionnaire's disease or HIV. There are two distinctions that make clear the difference between an intern in these sorts of circumstances and Engelhardt's indentured adolescents. First, physicians in such cases take all available precautions to minimize the risk to their own health, whether that be masking and gowning to minimize respiratory droplet contact or universal precautions to minimize contact with possibly infectious blood and other bodily fluids. Second, the responsibility for treating patients even in dangerous circumstances comes not from the indenture, but from the obligations inherent in the profession of medicine itself. An attending physician may not order an intern to risk her life, but the intern's (and attending's) profession can—and does—require it. By the analogy Engelhardt makes, it would have to be the "profession" of being an adolescent that obligates one to risk one's life, which is clearly not the case.

The implicit contract, if such it is, of obedience in exchange for support between parents and children surely does not grant more justification for the parents to make life-and-death decisions than is granted in these more clear cases of indenture. In any case, even if it were held that the implicit contract of indenture allows parents

---

[65] *Foundations*, 2nd ed., p. 156.

to make such choices, it seems reasonable that the child simply could reject the terms of indenture, surrendering any claim to future support and becoming free of any obligations of obedience. This again means that the contract for indenture, even if plausible, is not helpful, at least not in cases where extremely important issues are on the line. Of course, most young adolescents cannot feasibly leave their parents' homes in this fashion, but they could be removed from those homes by social services. This removal, which is ostensibly an imposition of governmental force against the parents, could be justified by the fact that a child's leaving the "indenture" would be a reasonable and prudent reaction for a person in such a circumstance.[66]

To assume that the relationship between adolescent children and their parents necessarily implies that parents may choose to refuse lifesaving health care for him is really not to have left the first (ownership) understanding of their relationship at all. It is true that each parent could choose for him- or herself to refuse a transfusion; it is also true that they may maintain control over their possessions, to the point of destroying them. But it does not follow that they ought to have that control over a fourteen-year-old child, even though they produced him and continue to feed, shelter, and clothe him.

In addition, the fact that Engelhardt's account implies that children may choose to enter into an indentured-servant-like relationship with their parents is, at best, confusing. Either children are capable of making such a contract, which would mean that they also are capable of making their own decisions and thus are persons in the strict sense, or they are not so capable and thus any such indenture could not be valid. Even implicit contracts can be valid only between persons in the strict sense; the problem here is the very question of whether—or, better, to what extent— adolescent humans are persons in the strict sense. Resolving the matter by appeal to a contract does not seem to be a resolution at all.

Nor can the concept of social personhood resolve this sort of case, though it is more interesting than in the case of an infant. Although an adolescent quite likely should be considered a social person—indeed, this is precisely the area of uncertainty addressed by Engelhardt's second justification for social personhood—it is far less clear just what assuming social personhood entails in this case. In the first case, it was relatively clear, if the social personhood construct worked at all, that the infant's social right to life should outweigh any presumptive right to freedom of choice of religion. But in the case of the 14-year-old, one cannot justifiably prefer one social right to the other on secular moral grounds, since the question of which right is more important is answerable only in contentful terms. An appeal to social personhood cannot help to resolve the issue either.

---

[66] And this point raises again the concerns noted in the above two cases about the inability to non-contentfully define "reasonable and prudent," and the difficulties in choosing an appropriate guardian for a non-competent person. No additional help for this resolution is given here, but the problems do complicate this case as well.

## Engelhardt's Principle of Intervention

Engelhardt responds to concerns such as these with an explicit principle delineating when and how one may employ force to intervene on behalf of a ward against the wishes of the ward's guardian, which is essentially what the doctors in Cases 2.2 and 2.3 would be doing were they to recommend overriding the parent's treatment preferences.[67] He recognizes that mature minors are often only partly "in their own possession" and thus can be partly emancipated. For some minors, then, as argued above, he holds that the scope of the indenture to the parents would not include the choice of medical treatment in life-risking circumstances, and intervention to provide that treatment would be justified.

Rather than resolving the above problem of not being able to determine the scope of any such contract, however, this principle runs squarely into it. How would one know whether any given adolescent is one who has taken these areas "partially into [his] possession by being morally responsible for such choices"?[68] When is it morally right to interfere in the parents' decision? Engelhardt's principle for intervention on behalf of a ward is:

> Out of consideration of violations of the principle of permission and/or the principle of beneficence as the principle of nonmalevolence, one may use... force to rescue a ward from a guardian in authority, if and only if:
>
> i.   The child asks for rescue, is competent, and the guardian's actions or omissions injure the body or mind of the ward to a degree significantly contrary to the best interests of the ward, as determined by the standard of the rescuer, and the rescuer pays any costs imposed upon the guardian; or
>
> ii.  The [guardian's][69] actions or omissions are malicious, that is, malevolent; or
>
> iii. The actions or omissions are contrary to agreements made with the ward before the ward became incompetent; or
>
> iv.  The actions of the guardian are such as are very likely to be interpreted as direct injuries by the ward and the ward is competent.[70]

The problem with this principle is that, though it is meant to apply to cases just like Case 2.3, it is not very helpful in doing so. The first or fourth criteria are the most applicable, but they both depend on the ward being competent, which, to a large extent, begs the crucial question. Regarding the first, if the ward were in fact competent, it already would be appropriate to employ force *via* a court order to support his decisions if his parents refused to honor them. Such a refusal would, in essence, be an imposition of unconsented-to force upon an innocent person by the parents, in which case the use of force to intervene is authorized already under Engelhardt's

---

[67] *Foundations*, 2nd ed., pp. 327–330.

[68] *Foundations*, 2nd ed., p. 327.

[69] Engelhardt (personal communication, July 2000) confirms that the text should read "guardian's" as written here.

[70] *Foundations*, 2nd ed., p. 328.

theory by the principle of permission itself.[71] The fourth criterion seems like the first, and perhaps is meant only to accommodate cases where a ward is, for example, temporarily unconscious, but would express herself as in the first criterion if she were able to do so. Both conditions require that the ward be competent to make these sorts of decisions. Since these two conditions of the principle only apply to children who currently are competent to make decisions about their well-being, they are not helpful in Case 2.2 and of extremely limited help in Case 2.3. How either condition ought to apply when a ward has "partially" brought himself into his own possession is left unclear.

So, the first and fourth criteria cannot help address this case, since the partial competence of the ward is a large part of the problem. Since the third criterion is not applicable here, the only part of the principle left that could be any help is the second criterion. In Cases 2.2 and 2.3, neither the parents nor the health care practitioners are malicious; they are all trying to do what is best for the patient. So, in the cases as written, the second criterion does not apply to either case.

But even in a case where the second criterion did apply, there are severe problems in any implementation of it. In Engelhardt's theory, there is no understanding of what beneficence, and thus maleficence, entails that can be delineated outside of a contentful moral view. There is no secularly defensible content to the concept of beneficence or maleficence. So one must ask on whose grounds the determination of maleficence should be made. If, in a given case, both the guardian and the presumed interferer agree that an action is malevolent, then it is simple to see how interference is justified, but how often is it going to occur that guardians will act toward their wards in ways they recognize to be malevolent? The problematic cases involve situations in which the observer sees maleficence while the guardian does not—e.g., in a case of female genital mutilation condoned by a cultural subset in society but not by others, or in Case 2.2. Without any ability to objectively understand harms and benefits outside of a particular moral viewpoint, which Engelhardt's theory does not and cannot provide, this criterion is severely weakened. The criterion cannot support or oppose intervention in any case without clarification and defense of "whose maleficence" matters, yet there is no secular moral ground on which one could defend such a clarification. The argument for this claim is essentially similar to the argument given for Engelhardt's inability to determine "whose ownership" matters. In either case, one would be defending a particular view of what is important—i.e., a contentful interpretation of the moral good.

The child in Case 2.3 is at an awkward and murky stage for Engelhardt (and, to be fair, for many other theorists as well.) The child is not fully able to make his own explicit decisions and contracts, because of his incomplete judgment abilities and his youth and inexperience, yet he surely is past the point where one could consider him an owned object in any reasonable sense. He is able to make certain decisions and has distinct preferences and desires, as well as hopes, fears, loves and needs.

---

[71] See *Foundations*, 2nd ed., p. 123. This would not be true if there were a contract for the child to submit in this case, but as argued above, no such implicit contract for decisions of this sort can be inferred.

He is a person in some important ways, which ought, by definition, to remove him from the pool of things that can be owned; but he is also, in this case, stipulated to be unable to make consistent decisions and to be a constant, full person.

So Engelhardt is left in a quandary. He can neither recommend that the doctor follow the parents' advice, nor recommend that she follow her own advice, because in each case she would be choosing a particular theory of the good for her patient, which is unacceptable. The individual in question cannot make a valid decision regarding his own theory of the good, and if the child were understood to be fully competent to make this decision, then there should be no conflict between parent and physician, but only a patient making his own decisions regarding his own health care. The doctor is left with no options that do not entail choice of a theory of the good for the patient. Not only does Engelhardt's theory give no advice as to which choice would be correct—which could be consistent with its admitted incompleteness—but it also gives no way to act so as to resolve the problem that does not act against its own major rule against using, or at least choosing the good for, others.

## The Limitations of Contracts

The fundamental limitation of Engelhardt's theory that all of these cases have in common is that, in each case, there is a party to the proposed action who is not an active, Engelhardtian/Kantian person. The principle of permission limits any discussion of secular morality to only that which addresses active, rational, adult persons; yet as the cases above show, many of the cases of moral conflict encountered in modern society and modern medicine involve participants who are not persons in that sense of the term. What these cases show are the limitations of using contract language—which refers to interactions and agreements between rational, autonomous persons of approximately equal power—to describe these medical interactions.

The problem that Engelhardt's theory has for characterizing relationships in medicine is simply that relationships involving persons of approximately equal power who are not in a specific dependent relationship are only one of many types of relationships persons have; and rarely are those sorts of relationships present in moral conflicts in medicine. Persons meet in medicine as moral strangers most commonly in ways other than as fully-grown equals. The contractarian account of morality that the principle of permission allows is insufficient to explain these relationships properly. Annette Baier has said, "we need a morality to guide us in our dealings with those who either cannot or should not achieve equality of power (animals, the ill, the dying, children while still young) with those with whom they have unavoidable and often intimate relationships."[72] Whether or not this is true for the whole of ethical thought, morality in medicine must be able to deal with these sorts

---

[72] Baier, Annette C. (1994). "Trust and Antitrust", in her *Moral Prejudices: Essays on Ethics.* Cambridge, MA: Harvard University Press, Chapter 6, p. 116.

of cases, for medicine deals daily with the ill and dying, as well as young children and even animals (in medical research, in any case).

Where, as in Case 2.3, there is a communal moral understanding underpinning one of the relationships in question, the principle of permission also inadequately addresses that understanding. The contractual analysis of the relationship between a parent and a child involves a shifting picture from complete dependence of the child on the parent, through a point of semi-equality in strength, often through to the parent becoming dependent upon the child. Never through the entire lifespan of a parent/child relationship is it well described by a contract. Through much of the early parts of the relationship, the young dependent child can give very little in immediate return for the protection and support granted by the parents, and is in any case in no position to be able to assess and agree to a contract. One could not argue that the relationship is a contract for future repayment (e.g., the parent supports the infant and child through her dependency period in exchange for the support of the child when both get older), arguing that grown children necessarily owe a debt that ought to be repaid in kind to their parents.[73] One would hardly argue, for example, that a parent had no responsibility toward the child simply because she knew that either she or the child, or both, would die well before the parent would need support in old age, or before the child would be of an age to provide support.[74] Any attempt to describe a relationship with such a radically different power dynamic, and such an intertwined relationship, in terms suited for comparisons between persons of approximately equal status will be severely limited.

The problem is not restricted to cases similar to Cases 2.1, 2.2, and 2.3, though that alone would provide a serious problem for any theory attempting to provide an ethics for use in medical situations, as these are common sources of problematic cases in medicine. Though there are cases where a moral problem arises when, for example, a competent ill person requests help from a competent other in committing suicide, many other cases in medicine involve less competent and less autonomous beings. When, as often happens in medicine, one encounters as a moral stranger an unconscious adult, an infant, a severely mentally retarded human, a research animal, etc., these are very often the sorts of cases and relationships that produce moral conflicts. For a secular morality being employed to provide a means other than force for resolving moral conflicts, it seems clear that an inability to accurately represent such situations is a severe flaw.

Yet that is not the whole of this concern with Engelhardt's theory. The larger problem is that libertarian language is poorly suited for modeling and analyzing the interactions between physicians and patients in nearly all medical interactions. The paradigmatic relationship for a libertarian interaction, an interaction between persons of approximately equal power who are not in a specific dependent relation-

---

[73] See English, Jane. (1979). "What Do Grown Children Owe their Parents?" from *Having Children: Philosophical and legal Reflections on Parenthood*, Onora O'Neill and William Ruddick, eds. New York: Oxford University Press.

[74] Baier, "Trust and Antitrust", pp. 109–110.

ship, is rarely the relationship encountered in health care. This is so even when an encounter involves a conscious, coherent, adult person. The fundamental relationship in medicine is the physician/patient relationship, which is inherently unequal and dependent with regard to three crucial elements: knowledge, power to act in the situation, and ability to debate or negotiate.

For most patients in such relationships, physicians have a monopoly on knowledge and power. They can make diagnoses and prescribe courses of treatment; the patients often neither have the specialized knowledge to diagnose the problem nor the access to the necessary drugs and/or equipment to treat it. Moreover, most patients are not only uneducated in medicine and their particular condition, but also are ill, frightened, or dying, which can put one in a mental state even less compatible with careful contract-making and permission-granting; to add to the inequity, patients are usually completely dependent upon health care professionals to alleviate these symptoms as well. Finally, the one source of power which all negotiators should have—the ability to leave negotiations if they are not going to one's liking—is not a reasonable move if one believes that one's life is on the line.

Consider the case of "Iphigenia Jones," as described by Jay Katz.[75] An "attractive, then-21-year-old, single woman" was diagnosed with breast cancer after a biopsy determined that a lump was indeed malignant. Her physician told her that mastectomy, possibly radical, was the only option for treatment, while other physicians at the same hospital performed and might have recommended a lumpectomy followed by radiotherapy. She accepted, not knowing that there were any options. The evening before the surgery was scheduled, the doctor went to her, full of misgivings, to tell her of the other options, though he still fully believed mastectomy was the best option for her. It was only the fact that the operation was to be performed on someone so young and attractive that made the doctor consider this last-minute approach to the discussion; as it turned out, Ms. Jones did prefer the less mutilating procedure and eventually chose lumpectomy and radiotherapy.

One could rightly note that, without this additional information, Ms. Jones was unable to give adequate informed consent to her procedure; but the point of the example is more fundamental than that. Whether or not the physician was right to present these options to Ms. Jones, or whether he ought to have presented them long before, the point is that the physician was in complete control of whether or not to present them. Ms. Jones' ability to make a good decision about her treatment was controlled by her physician's willingness, or lack thereof, to reveal the information that he had about her options. He had all the relevant knowledge, and only revealed it at the last moment for relatively shaky reasons. The knowledge was his, and the choice to reveal it was his. Ms. Jones had little to no knowledge of her disease or its treatments other than what he gave her, which is not surprising for a young person. She also had very little power to reasonably influence or choose her treatment until the physician gave that power to her, which he almost did not. It is true that she could have chosen to refuse the mastectomy before knowing that there were other

---

[75] Katz, Jay. (1984). *The Silent World of Doctor and Patient*. New York: The Free Press, pp. 90–93.

treatment options, but this would not have been reasonable for her when she believed that mastectomy was her only chance for survival. Likewise, she could have gone to another doctor, but would have had no good reason to if she didn't know there were other options. Though she did, in the end, make a decision based on her own desires and values with the help of the information her physician granted her, her ability to do this was controlled almost entirely by her doctor. The physician controlled the knowledge and the power to act on that knowledge, because their relationship was unequal.

It is true that physicians were never the only possible source of information about a patient, and this is more so now. An internet-savvy patient or relative can find a wealth of information, some of it good, about breast cancer and a fair number of other more common diseases. So today's patients need not be as bad off in their interactions as Ms. Jones was. Yet even armed with information, a patient is still at the mercy of the physician to know what of the information is any good— e.g., Laetrile is not touted on better cancer-related websites, but anecdotal cases are definitely extant on the Web—and what of that information applies to her particular case. Patients also remain, by the restrictions of the medical system, under control of physicians in order to receive any treatment. The situation is better now than when Ms. Jones was treated, but not fundamentally different.

One could argue that the physician ought to have given her this additional information, and much sooner, and such a conclusion would be correct. But that obligation to provide information fails to mesh with the paradigm of interaction in Engelhardt's theory. The point is not that the physician acted wrongly in not telling her or rightly in telling her, but that Engelhardtian terminology provides little to describe the subtleties of power and information differentials that make all the difference in judging the rightness of any particular part of the interaction. The model for contract theory is the free market, and the paradigmatic decisions and contracts made in that arena: two or more persons, of relatively equal power and understanding, enter into agreements in an attempt to maximize their own benefit. With this analogy one can see how ill suited such a theory is for modeling medical interactions and decisions. At almost no time did Ms. Jones's interactions with her physician resemble that model. What good businessperson would ever willingly enter into contract negotiations where the other contractor was the only one who knew what was going on and what was at stake, where the other contractor was the only one with any ability to act—or access to the tools necessary to act—when and if she chose to reveal what was going on, and where the first businessperson was also feeling ill and frightened? Yet this is exactly the way that frightened patients like Ms. Jones must put themselves into the hands of their physicians, even if they try to even the balance a bit by informing themselves about their conditions.

Even when one free-market contractor knows she has less to offer than the other, and is thus unequal in power in that sense, she would still seek to avoid the conditions of fear, ignorance, and illness, as they would make the negotiations worse for her. An agreement could come out of negotiations under these conditions, of course, but no one would expect it to be anything like the agreement that would come out of the same negotiations with two debaters with similar knowledge, access and ability to act, and clear-headedness. The agreement would be expected to be quite unfair, and it would be surprising if the more powerful contractor did not push her advantage to make it so. This is not the sort of interaction, however, that we would seek for physicians and patients. The relation of the alternate possibilities to Ms. Jones was entirely appropriate, as the physician ought to care not only about his own benefit but also that of his patient. It was inappropriate only because it was done as a last-minute "attack of conscience." Such an "attack" is utterly unmotivated by a free-market-based desire to maximize self-benefit. It is not surprising that the beneficent parental figure had been the role model of the physician for so many years, because it fits the interactions between patient and physician so much better than does the model of approximately equal contractors.[76]

In cases like Ms. Jones's, particularly in the era of more managed care, patients often also have minimal alternative options; though some with good health insurance may choose another doctor if they have the time and ability, they often cannot do so in an informed fashion. In order to know whether one ought to change doctors, one would need to know the information that, in this relationship, the patient usually does not have, such as specialized medical knowledge about his condition and a new doctor's capacity to treat that condition as compared to other doctors. Patients are also often restricted in their ability to leave negotiations if they are not going well. Many patients in critical situations do not have the option of leaving, even if they did have the requisite knowledge, simply because the cost to their health or lives is too great to risk or, in many cases, fully comprehend at the time. Insurance companies are also often unwilling to pay for treatment if a patient record indicates that they have left care against medical advice. These are common instances of the physician/patient relationship, especially in the sorts of cases that may be difficult moral conflicts: drastic power and knowledge differentials, dependency, some lack of capacity for full autonomy.

All of these features, which are all too common in the modern medical landscape, serve to make the normal interactions in medicine less and less like contract negotiations. The physician/patient relationship is very poorly explained by traditional contract theory, framed as it is on interactions between two autonomous persons seeking only their self-interest in a situation where they have reasonably similar knowledge and at least the amount of control over the debate granted by an ability to walk out if the proceedings are not going to one's liking.

So, the problem with Engelhardt's theory is that it cannot accurately model the kinds of interactions that are frequently found in bioethics, and thus the solutions to the problems that it creates will be based on models of the actual situations that are not very close to the actual situation. It can describe the rough outlines of the problems well enough, but will not be able to get at the nuances that often mean the difference between a simple moral problem and a difficult one. A theory of bioethics

---

[76] It is a truism in hospitals, etc., that physicians make the worst patients, possibly because they do not fit the mold of the unknowing, weaker patient needing the physician as his conduit of access to the tools of treatment.

incapable of describing the basic relationship in medicine seems poorly suited to the task of resolving moral problems in bioethics.

## The Second Argument—A More Minimal Ethical Grammar

Still, Engelhardt could, at this point, argue that even if all of the above is true, this is simply one of the tragic results of being limited to secular ethical language and justifications. He could argue that, though the principle of permission is limited, there is no other choice but to accept it in order to have a non-question-begging set of conditions for a common, secular moral authority.[77] It is a necessary principle—and the only necessary principle—for secular ethical debate or for attaining secular moral authority. He could conclude that it is unfortunate that secular moral justification cannot do everything one might like, but nevertheless the principle of permission is all that secular moral authority can authorize, and so if one wants to have a means other than force to resolve moral conflicts in a secular society, permission is all that is left.

Such a reply would be consistent with Engelhardt's defense of the principle of permission.[78] It is, however, mistaken. It argues that the principle of permission is appropriate for secular use because it is both content-free and necessary for the use of moral concepts and language in a secular environment; neither of these claims is true. Recall that Engelhardt's argument for the moral validity of the principle of permission began by an argument by elimination: if conversion or rational agreement don't exist, as they don't for moral strangers, then we can only employ force or agreement to resolve issues. Force is non-moral, so the only grounds for a moral justification is agreement—or, more precisely, the principle of permission, which also provides grounds for resolving matters when there is no agreement by holding that one must (with a few exceptions) leave others alone to pursue their own notion of the good. Only the principle of permission can provide grounds for moral concepts between moral strangers. It is the fail-safe, default position that is our only option for moral resolution between moral strangers.

If, however, there is another way of providing for moral language and concepts between moral strangers, then the principle of permission is not the only option for resolving conflicts in a moral fashion in a secular sphere. (It may be true that this principle is necessary for a particularly useful secular morality, but that is not relevant to this argument.) If one can coherently apply the moral terms of praise and blame without the principle of permission, and without a more contentful principle, then the principle of permission is not the *only possible* means of successfully appealing to moral authority in the secular sphere. Further, as will be shown, the principle of permission actually involves a substantive moral assumption, which

---

[77] Engelhardt, "The Foundations of Bioethics and Secular Humanism: Why is there No Canonical Moral Content?", p. 271.

[78] See, e.g., *Foundations*, 2nd ed., pp. 67–71.

runs counter to the Engelhardtian agenda of procedural and so content-free secular morality. It is not what he would call a "content-full" theory, but neither is it free of content.

Even in the face of this, one could still argue that the principle of permission is an appropriate principle to add when trying to make a secular system functional. It may not be content-free, this line of thinking would hold, but it is appropriately "content-thin." If a content-free morality is not possible, perhaps an Engelhardtian could argue along these lines that the best secular principle is the principle of permission. This argument, too, will not succeed. If, *contra* Engelhardt, the principle of permission is not required for a secular moral language of praise and blame, then there is little reason, given the above criticisms, to hold that this particular principle ought to be taken as the primary principle of secular morality. This is not immediately obvious—the secular moral system that is available without the principle of permission is even poorer than Engelhardt's admittedly lacking system. But once one has surrendered the hope of a content-free system, there is little reason to add the principle of permission as one's chosen contentful principle. If one adds to a moral system something more than that necessary for an appeal to secular moral authority, then one must ask the question, "What further assumption should be made, and why?" If the goal is to make that appeal to moral authority actually useful for resolving moral conflicts, the principle of permission does not fulfill that goal, and hence Engelhardt's theory will not serve as the best means to resolve moral problems in a pluralistic society.

## *Wanted: Moral Analysis*

Engelhardt wants to argue that the principle of permission is the only possible means of achieving moral conflict resolution in a secular society. That it is a *moral* resolution is important. Persons may, of course, choose to enter into an agreement to follow the principle of permission when they meet as relative moral strangers. Indeed, one reading of the *Foundations* seems to suggest that this is just what is done in such a situation in order to resolve problems, holding that the authority of secular morality is, and can only be, created by persons choosing to accept permission as a means of resolving moral conflicts.[79] But there seems to be more to Engelhardt's account than simply a choice by some persons (but perhaps not others) to enter into a particular game, which would then entail following the rules of the game. The principle of permission is not treated by Engelhardt as merely the rule of one sort of possible "interaction game" that one could choose to follow if one wanted to. It could function as a rule in such a game, of course, but it is also treated as more

---

[79] Aulisio, Mark P. (1998). "The Foundations of Bioethics: Contingency and Relevance." *The Journal of Medicine and Philosophy* 23: 428–438, esp. p. 432. See also Wildes, Kevin Wm. (1997). "Engelhardt's Communitarian Ethics: The Hidden Assumptions", in *Reading Engelhardt: Essays on the Thought of H. Tristram Engelhardt, Jr.*, Brendan P. Minogue, Gabriel Palmer-Fernandez, and James E. Reagan, eds. Boston, MA: Kluwer Academic Publishers, pp. 77–93.

than that: it is supposed to be descriptive of morality, which is understood to be a particular sort of interaction. One can praise or blame in the context of different sets of rules, such as the rules of baseball ("You should not have run outside of the base path on your way to first base") or the rules of etiquette ("One ought not belch in front of the Queen"). Morality includes not just the capacity to praise and blame, but the capacity to praise and blame in a particular way. The evaluation possible *via* the principle of permission is meant to be specifically *moral* praise and blame, which is intended by Engelhardt to be something more than an analysis of whether you have merely broken an arbitrary, albeit freely chosen, rule.

Even if followed only because freely chosen, the principle of permission is made distinct from other sets of rules by the labeling of its rules as moral. As well, the principle of permission is, on Engelhardt's account, a necessary principle for secular moral interaction, and it must be both necessary and moral to fulfill its role in Engelhardt's system.

But is the principle of permission both necessary and moral? It does have what Engelhardt calls a "family resemblance" to the various forms of moral reasoning that have been held to be true throughout human history. Various differing sources of morality have been proposed during that history—reason has been thought to be one; religious perception and intuitive perception of moral truths have been two others. These justifications, and the moral systems that have come out of them, differ significantly but share some basic similarities, or resemblances.

This resemblance cannot be merely a similarity in form, at least partly because there is little similarity in form between the various classical types of moral theory and moral systems of conflict resolution that Engelhardt discusses.[80] For example, reason-based moral theorists and theological intuitionists hold radically different bases of morality to be true, and basic differences in the source of morality as well. The two address similar questions, but often in very different ways.

Yet there are at least three ways in which they resemble each other, and thus in which a family resemblance could be located. First, they address questions and problems generally recognized as moral. Each allows persons to address certain kinds of central issues (regarding acts like killing, truth-telling, and the like) that have been traditionally recognized as moral questions. A moral theory, whatever else it might be, needs to address these sorts of questions in order to be a moral theory. Though they are directives, the rules of baseball are not moral rules; the rules of truth-telling are. Second, moral theories grant a means for praise and blame in a specifically moral fashion, and often make it clear what the appropriate moral punishments for blameworthiness are. Third, they generally share a method of reasoning from basic moral truths to more specific ones; whether the basic moral truth is perceived by pure reason, by God's grace, or by an intuitive sense, the understanding of what follows from that moral truth generally is developed by logical progression from basic truth to more developed truth. That is, they all use basic logical reasoning.

---

[80] *Foundations*, 2nd ed., pp. 40–64.

Any moral system, including that based on the principle of permission, Engelhardt argues, grants a means to address these problems that are classically understood to be moral problems. It allows one to place blame and praise persons on grounds that are specifically moral, while employing good basic logical reasoning; and the principle of permission is understood as a specifically moral principle, because of this resemblance to classical forms of moral praise and blame.[81] On these grounds, it seems that the principle of permission is a moral principle.

However, if this is the family that a moral theory must resemble, it is not clear that the principle of permission is necessary in order to make non-question-begging sense out of moral praise and blame in a secular setting. Engelhardt argues that the rejection of force *against innocent others* is the minimum necessary premise for the possibility of morality—that, if one wishes to avoid foisting moral content on others, but still wants to be able to speak morally, this is the "only game in town"[82]— but this is in fact incorrect. One can allow for praise and blame in a secular sense without appeal to the principle of permission, and more precisely without rejecting the use of force against innocent others as Engelhardt does. Therefore, the principle of permission cannot be justified as the sole possible grounds for a secular morality.

The family resemblance model, or something similar to it, can be a good means of distinguishing between moral and non-moral systems. In order to seek moral responses to moral problems, some means of determining whether a system is moral is needed. The family resemblance model is perhaps not perfect. For example, by being focused on historical features of moral theories, it is necessarily conservative. There could be moral questions which have historically been ignored, or which could not have accounted for historically because of lack of awareness of important facts of the world or lack of ability to bring about some new potentially moral situation (e.g., the moral status of early embryos or of human clones.) It may also preserve as "moral issues" issues that are not moral, though they have been historically thought so for idiosyncratic reasons. Nevertheless, the family resemblance model can provide an important baseline distinction between moral and non-moral systems, and though it may not be complete, it is a good starting point. The family resemblance model is capable of eliminating at least some explicitly non-moral systems from being contenders for providing a moral response to problematic cases.

The principle of permission should therefore meet the criterion of morality set by the family resemblance model, even if it can only do so weakly. It could not account for the traditionally understood importance of altruism and generosity, for example. Since it is such a sparse moral principle, however, perhaps it is inappropriate to demand accordance with all of the features of family resemblance. After all, it does resemble the family of moral theories with regard to a very important feature. It defines the appropriate limits of force against another person, and does do very clearly and precisely. Perhaps this is its area of strongest resemblance to morality.

---

[81] Engelhardt, H. Tristram, Jr., personal conversation, July 2000.

[82] Aulisio, "Contingency and Relevance." p. 433; see *Foundations*, 2nd ed., pp. 67–71.

If Engelhardt's theory is going to hang its moral hat, so to speak, on the rejection of force against innocent others, it is worth discussing that rejection. As will be seen, the rejection of force is actually a problem for Engelhardt instead of a strength.

## *The Rejection of Force*

The principle of permission is defended by means of elimination of all other possible means of conflict resolution—it is a "last resort" principle that is accepted because some means of moral conflict resolution is needed and there are no other options.[83] As a part of this justification of the principle of permission, Engelhardt rejects force as a possible moral means of justifying a conflict resolution.[84] However, his argument for why force *against innocent others* is not a moral means of conflict resolution is insufficient. I argue that Engelhardt presumes, rather than proves, the moral preference for agreement; on the grounds that he lays out for secular moral debate, he cannot show that agreement and permission are morally superior to willful imposition of a view on an innocent other. His rejection of unconsented-to force is a contentful moral claim that cannot serve as part of the basis for a secular pluralist morality.

That this is a contentful claim can be seen when one asks precisely why the appeal to secular ethics means one must reject force as a means to resolve moral disputes,[85] e.g., why is it that (peaceful) tolerance is the central virtue of the secular society?[86] This is the central question and the primary challenge to this secular ethical theory; Engelhardt has not adequately justified this claim.

In fact, Engelhardt's theory works to the extent that it does only because of a presumption in favor of consent and agreement which, although it is reasonable, is a contentful moral presumption rather than a necessary precursor of any moral claim. If this is so, then a content-free, procedural theory cannot achieve even the moderate goals of minimal conflict resolution that Engelhardt claims for it. Without that contentful claim, Engelhardt cannot make the principle of permission, defended on the grounds of the moral rightness of agreement and the general moral wrongness of force, the central pillar of a theory of moral conflict resolution that can be defended by and to all members of a secular, pluralistic society.

In order to prove this, it must be shown that one cannot, on secular moral grounds, defend agreement as a specifically *moral* method of conflict resolution. The rejection of force against innocent others is common to most moral theories, in some form or another, but cannot be defended on secular moral grounds because it

---

[83] See *Foundations*, 2nd ed., pp. 67 ff.

[84] Engelhardt does not reject all uses of force; in fact, he explicitly authorizes it for enforcing contracts, etc. But he does reject force as a means of achieving moral authority. See *Foundations*, 2nd ed., p. 67.

[85] *Foundations*, 2nd ed., p. 67.

[86] *Foundations*, 2nd ed., pp. 419–420.

involves a contentful assumption about what morality entails. Once this is shown, it can be shown that there is a competing theory of secular morality, at least as justified as Engelhardt's, which has the same family resemblance claim to morality as does the principle of permission.

This is not, of course, a new challenge. It has been asked of Engelhardt's work at least since the first edition of *The Foundations of Bioethics*.[87] This is often presented as a challenge that Engelhardt does not address; but Engelhardt does respond to this, although ultimately unsuccessfully. He provides two lines of argument for why force must be rejected and the principle of permission accepted in a secular bioethics: the need for intellectual authority and the need for moral praise and blame. Neither will serve to necessitate the principle of permission.

## *The Argument from Intellectual Authority*

Why ought one to prefer rational agreement to coercion, compulsion, authority, or violence in making one's moral points? In many cases, the latter can be significantly more effective, so why is the former preferable; or, more to the point, why is the former thought moral and the latter immoral? Engelhardt has stated that a "goal of ethics is to determine when force can be justified,"[88] which is consistent with the family resemblance model above. He holds that force against innocent others is unjustifiable, but he cannot, on the grounds he allows himself, show that force against innocent others cannot be justified. None of the approaches available to a secular moral argument can work to prove this: it is not part of the definition of morality, nor is it necessary in order for one to make moral claims, to reject force in the way that the principle of permission does. Nor can one argue that the rejection of force is the only way to make a secular morality have any kind of resemblance to other kinds of morality, which could be used to explain why a secular conflict resolution method ought to be considered the only *moral* method. In fact, as shown below, on secular moral grounds one cannot argue that the rejection of force against innocent others is morally superior to the implementation of force. None of the arguments for the rejection of force on the level mandated by the principle of permission can be successful on secular moral grounds. The rejection of force as a means of justifying a moral claim can only be defended in the context of a contentful view, which, it is argued below, is what Engelhardt has provided.

Perhaps the initial secular argument for why permission is preferred is that it is obvious that agreement is morally preferable to undesired force, and that tolerance is *prima facie* better than intolerance. For most liberal western readers this is a normal starting place for moral reflection, and thus it does not raise our hackles

---

[87] Loewy, Erich H. (1987). "Not By Reason Alone: A Review of H. Tristram Engelhardt, Foundations of Bioethics." *Journal of Medical Humanities and Bioethics* 8(1): 67–72. For a more recent phrasing of the question, see Nelson, "Everything Includes Itself in Power", pp. 20–23.

[88] *Foundations*, 2nd ed., p. 67.

when Engelhardt makes it the starting point of his theory—but it should, as contentful claims are the source of most readers' beliefs that tolerance and minimizing violence are morally preferable.

Engelhardt initially explains why force is an inappropriate means of conflict resolution in a secular, pluralistic context by appeal to the authority by which a problem is solved. In order to intellectually resolve conflicts one must seek some method other than force because:

> Resolution by force carries no intellectual authority either with regard to (1) which viewpoint is correct, or (2) whether the correct viewpoint *may* be imposed by force.[89]

If intellectual authority is understood as meaning giving good, rational, or rationally compelling reasons for agreement to a particular claim, then it is clear that force does not carry intellectual authority. Force, as Engelhardt notes, is simply force, and intellectual authority so defined is gained by the successful justification of a claim. Force does not carry any intellectual, justificatory authority with regard to determining which moral position is correct or whether that position may be imposed by force.

True enough, but what is shown by this? Or, to phrase the same question differently, why does one care whether something has intellectual authority? The reason has to be that intellectual authority is important for something to be a *moral* claim, which is why it is a part of the family resemblance model. Intellectual authority is obviously desired for the justification of claims and actions because that is what makes a reason to believe a claim a good one.[90] Force, by dint of its lack of intellectual authority, is therefore prohibited from being a good justification for a moral claim—it seems fairly clear that "because I will punch you in the nose" is not a sufficiently good response to a request for a justification.[91]

This is sufficient to show that force is not an acceptable ground for thinking one particular viewpoint is correct. However, the justification of the principle of permission also requires that one may not impose the correct viewpoint by force.[92] The arguments provided so far have not shown that it cannot be morally correct to use force to modify other persons' beliefs. One could still plausibly argue that force is acceptable, in at least some situations, as a tool to get persons to believe what one takes to be morally correct or at least to act in accordance with it. Force has no value

---

[89] *Foundations*, 2nd ed., p. 71 (italics in original).

[90] If it were not, then any justification for a claim would suffice: "Performing abortions is wrong (or right) because the sky is blue." The appeal to intellectual authority is in opposition to moral theories that have no role for rational justification, if any such theories exist; Engelhardt is discussing only those in which rational justification plays a vital justificatory role.

[91] Though it is typically sufficient to end the conversation.

[92] Strictly speaking, the second half of the quote means "force carries no intellectual authority to show that one may or may not impose the 'correct' viewpoint on another." This is, of course, true— force carries no intellectual authority to prove any claim whatsoever—but the more interesting question, which seems to be Engelhardt's point and must be answered for his theory to work as it does, is whether *anything* can show on secular grounds whether or not force can be used to impose the correct viewpoint on another.

as a reason, but the use of force could still be justified by other reasons; although force itself provides no intellectual authority, it may yet be moral to employ force if a reason with intellectual authority does justify it. For example, one could argue that one might justifiably use force to be able to treat the infant child of parents refusing treatment in Case 2.2—whether that be state-sponsored force through a court order or simply the physician's individual refusal to cease treatment—because one ought to save the infant's life. The force does not provide the justification; the rationale behind the force does.

So, that force itself has no intellectual authority gives, as yet, no reason to think that the correct viewpoint can never be imposed by force. But if this cannot be shown, then violence, or the threat of violence, against innocents can be justified by moral reasons, despite the clear violation of the principle of permission that this entails. Such violence seems to be perfectly consistent with the appeal to intellectual authority Engelhardt suggests is required for morality. Consider the following purportedly moral argument:

### Argument for Force

(1) The ending of human lives is a wrong, and it is an additive wrong—that is, the more lives lost, the worse an action is.
(2) Abortion is wrong because it ends human lives that would not have otherwise ended at that time.
(3) If Dr. Y survives, he will continue to perform abortions for some time.
(4) If Dr. Y is killed, he will perform no further abortions.
(5) If Dr. Y is killed for performing abortions, and people know that, and a credible threat to continue such killing is made, fewer people will want to perform abortions.
(6) Thus, if I kill Dr. Y, fewer abortions will be performed.
(7) Thus, I should publicly kill Dr. Y, as this will minimize the number of human lives lost.

If certain factual assumptions are correct, then the (otherwise?) plausible moral belief in (1) entails that one ought to kill Dr. Y. Even if the argument eventually fails because of flawed factual premises about, for example, how people will respond to such a killing, this does not make the argument in principle wrong on moral grounds. That would simply mean it would not be a good idea in this particular context to kill Dr. Y, although in a different factual context—if, for example, Dr. Y is the only doctor in a particular area who will perform abortions—it could be. The argument seems to justify killing, or at least kidnapping or incapacitating or otherwise violently stopping, Dr. Y. Although the Argument for Force reaches a conclusion that strikes most as unacceptable, it is at least *prima facie* a perfectly plausible and valid argument that is perfectly consistent with the logical requirements of intellectual authority for something to be a moral claim. It seems to follow an acceptable and common "grammar" of moral discourse, addresses a moral issue, and it allows for praise and blame, which are the signs of moral family resemblance noted above. But what the killer of Dr. Y is claiming to be doing is imposing, by force, compliance with what he understands to be the best possible outcome, and the correct moral viewpoint. The commitment to seeking intellectual authority cannot show this argument to be an unacceptable one; but the principle of permission requires that it be unacceptable. Consequently, the appeal to intellectual authority cannot justify the principle of permission.

It should also be noted that an appeal to intellectual authority cannot directly suffice to require one to prefer agreement over force either. Neither force nor agreement carries any moral or intellectual authority, per se. Despite its ability to peacefully end moral (and other) conflicts, mere agreement actually carries no more intellectual authority than does mere force—although it will usually allow people to settle a conflict, "Everyone agrees to this," is itself no more a good reason to believe a claim than is, "Do this or I'll punch you." Mutual agreement often serves as a proxy for good reasons, but it is only a legitimate proxy if those who agree truly have good reasons for their agreed-upon beliefs. So, on the grounds of intellectual authority, one cannot appeal to agreement as morally superior to force as support for non-violent solutions needed to justify the principle of permission. Rather, one can only appeal to the provision of good reasons, which can be given in particular cases for either agreement or force.

If the argument from intellectual authority cannot successfully prove on secular grounds that it is wrong, in a morally relevant sense, to use force to impose what one understands—on contentful and thus unshared grounds—to be a morally correct view, then the use of force on unconsenting others could still be held to be a potentially morally legitimate means of resolving at least some moral conflicts. The principle of permission would not necessarily be morally preferable to a "principle of force," at least in some cases. Unless this "principle of force" can be rejected, secular reasoning amongst moral strangers will be virtually impotent to remove force, coercion, and violence from the realm of the moral. Yet this is what Engelhardt's theory needs to do in order to defend the principle of permission.

Engelhardt does argue that the project of ethics is an appeal to something other than force as a means of resolving moral disputes.[93] His theory, as it is actually developed, requires not only that force not be appealed to as a source of moral justification, but also that unconsented-to force is itself always unacceptable if that force is not justifiable by the standards of secular morality. This is true even if use of that force is justified on grounds that do carry moral authority on an individual's own contentful moral theory. As Engelhardt describes it, unconsented-to violence to an innocent other is always blameworthy, even in difficult or tragic cases.[94] If this response can be justified, the Argument for Force is unjustifiable. However, as argued above, intellectually valid arguments can authorize the use of force even against innocent others; the argument from intellectual authority cannot eliminate

---

[93] *Foundations*, 2nd ed., pp. 67–69.

[94] *Foundations*, 2nd ed., p. 130.

this authorization. Another ground for rejecting violence against innocent others must be found in order to justify the principle of permission.

## *The Argument from Praise and Blame*

The response Engelhardt would give to this challenge is that tolerant non-violence, including both rejecting force as a means of justifying a moral claim and rejecting violence in all cases against unconsenting others, though not necessarily rejecting other applications of force, is the minimum necessary starting point for moral discourse between moral strangers. If one is to have a general secular ethics, one must accept the principle of permission:

> By appealing to ethics as a means for peaceably negotiating moral disputes, one discloses as a necessary and sufficient condition (sufficient when combined with the decision to collaborate) for a general secular ethics the requirement to respect the freedom of the participants in a moral controversy... as a basis for common moral authority....[95]

The point is that the principle of permission is the only means left to retain some capacity for attaining secular moral authority—a capacity for moral praise and blame—once one accepts the inability of contentful theories to be justifiable to all rational persons in a pluralistic society. The preference for peaceable resolution, without an appeal to force against unwilling others, is already included in the concept of moral authority. Yet, as a justification for this inclusion, Engelhardt only offers:

> If one is interested in collaborating with moral authority in the face of moral disagreements *without fundamental recourse to force*, then one must accept agreement among members of the controversy or peaceable negotiation as the means for resolving concrete moral controversies.[96]

In other words, the preference for peaceable resolution over violent resolution is simply the basic assumption necessary in order to accept secular moral authority. This will not, of course, convince Dr. Y's assailant that he must not use force; he has already accepted that force may be used in some cases to further the good, and thus rejected the basic assumption that one should collaborate without recourse to force. Engelhardt admits that zealots will not be convinced, but holds nonetheless that "the moral account justified [by his arguments] provides a secular moral warrant to authorize coercive force to protect persons when acting peaceably...."[97] The "zealot" who refuses to follow the principle of permission "loses the possibility of intersubjective collaboration in the face of [confronting] moral strangers."[98]

---

[95] *Foundations*, 2nd ed., p. 69.

[96] *Foundations*, 2nd ed., p. 68. Italics added.

[97] *Foundations*, 2nd ed., p. 11.

[98] Engelhardt, "The Foundations of Bioethics and Secular Humanism: Why is there No Canonical Moral Content?", p. 265.

True, the principle of permission provides a warrant for the use of force; but so do many other moral and non-moral accounts. Why use this account, rather than another? The argument seems to be that this is the minimum, the basic rules of the "grammar" of morality that allow for one to employ the concepts of moral praise and blame, with no presumption of particular content. It "provides the categorical possibility of being able to think a moral structure,"[99] or, put another way, it is the basic rule of the system of morality, without which secular morality cannot exist.[100]

This justification of the principle of permission begs a very serious question. It is true that the principle of permission is the basic rule of a particular system of inter-action. Those who accept this basic rule can function in that system, and those who do not cannot; but why is this system the one that gets the label of "morality?" This is the central question for Engelhardt. This system allows for praise and blame, but so do many other systems that are presumably non-moral. Persons who are moral strangers, except insofar as they want to resolve problems peacefully, could join together in a system of rules under the principle of permission, but persons who are moral strangers, except insofar that they believe that one should sometimes resolve conflicts by force, could also join together in a system of rules. Following Engel-hardt, the former group can call their system of rules moral, while the latter cannot; but the grounds for this are not content-free. It appears that the only reason the former can be seen to be morally superior to the latter is by a basic, fundamen-tal, and morally contentful assumption that peaceable resolutions are morally better than forceful ones. Engelhardt is including content in his theory—a minimal and reasonable content, but content nonetheless.

In an attempt to preserve the claim that the principle of permission is part of the necessary grammar of morality, one might argue that "resolving" moral problems entails a peaceful solution. In the definition of a resolution of morally problematic cases in the previous chapter, there is a good reason to think that resolution of prob-lems is necessarily a peaceful procedure, because "resolution" means presenting an option as the right one to pursue and justifying it to others in a way that they can recognize as valid. Therefore, "violent resolution" would appear to be an oxymoron.

It is true that such resolutions would generally be peaceful. However, the defini-tion of resolution herein is insufficient to defend the principle of permission. One cannot justify a claim by an appeal to violence and resolve a problem under this def-inition. But the principle of permission holds not only that, but also that one cannot successfully justify the use of force against innocent others; this is not contradicted by the definition of resolution here. "Violent resolution" is oxymoronic when what is meant by that is a resolution justified by violence, but the principle of permission demands more than this. It denies the possibility of any justified use of force against innocent others. This claim cannot be justified on the grounds Engelhardt accepts.

---

[99] Ibid., pp. 264–265.

[100] See, e.g., Engelhardt, H. Tristram, Jr. (1988). "Foundations, Persons, and the Battle for the Millennium." *Journal of Medicine and Philosophy* 13: 387–391.

Another response might depend on the non-moral nature of forced action. It might be argued that the use of force can be shown to be morally inferior to the principle of permission in Engelhardt's theory because the use of force does not allow for praise and blame, at least with regard to one who is forced into a given action. If we are not free to act, we are not responsible (and thus neither blameworthy nor praiseworthy) for our actions. A person being forced to perform or not perform some action cannot be moral or immoral in that behavior, as force restricts the autonomy necessary for moral assessment. The preservation of autonomous behavior is therefore necessary for morality and, on these grounds, the principle of permission could be thought superior to a principle of force. Agreement, even grudging agreement, preserves autonomy; force does not.[101]

Though an interesting defense of the morality of peaceable conflict resolution, this will not suffice to show the principle of permission as morally superior to the use of force to resolve moral conflicts between moral strangers. The principle of permission does not prohibit all non-agreed-to uses of force; rather, it authorizes the use of force in some cases. When a person has unconsented-to force, or unconsented-to malevolent acts, imposed upon them, under the principle of permission individuals are allowed, and a government may even be obliged, to respond with force to prevent such an imposition.[102] If one notes that forced action is non-moral, this force would prevent the imposer from acting morally—no praise would be accorded to him for his refraining from his forceful or malevolent act. Indeed, no praise would be appropriate if one refrained from malevolence simply to avoid punitive or defensive opposing force. So while it is plausible to argue that force denies the possibility of morality, this does not prove the principle of permission to be superior to the use of force; the principle of permission authorizes and even requires the use of force in some cases.

Such an appeal would also not show the principle of permission necessarily to be superior to other principles guiding the appropriate use of force. Non-autonomous action is non-moral, by this definition. For this reason, autonomy is valued morally, since it is a precondition for moral action; perhaps this is a non-contentful claim that can be employed to justify actions when interacting with moral strangers. (It may well not be non-contentful, as one could argue that, for example, minimizing autonomy would minimize the number of morally blameworthy acts, which might be valuable in variety of contentful theories.) But even if it is, it isn't clear that this recognition argues for the principle of permission. It might, instead, argue for a "principle of non-intervention," wherein one is prohibited from employing any force, even defensive or punitive, on the grounds that one's employment of force prevents actions that can be morally evaluated. Or, with a slightly different contentful set of values, one might argue for a consequentialist "principle of autonomy

---

[101] I am grateful to an anonymous reviewer who raised this point.

[102] See, e.g., *Foundations*, 2nd ed., pp. 108–110.

maximization," wherein force on others is justified and obligatory if, and only if, it will lead to the greatest number of actions taken autonomously. Even if, as might be argued, that consequentialist principle turned out to be the principle of permission, the justification for it would be radically different from Engelhardt's. So though there is an argument to be made here about the importance of autonomous actions for morality, such an argument is not sufficient, especially for an Engelhardtian who would not wish to appeal to consequentialist grounds, to justify the principle of permission as the sole appropriate moral principle for use with moral strangers.

If the foregoing is correct, then the principle of permission is not content-free. Still, one might argue, as some have,[103] that though the principle of permission is not completely free of content, Engelhardt is introducing only a minimal amount of moral content by this principle in order to achieve the desired end of being able to attribute moral praise and blame. Even if this is Engelhardt's goal—which he denies[104]—the principle of permission is not in fact the minimal amount of content necessary to enable secular praise and blame; therefore, even were this his aim, he is unacceptably importing too much content.

If one reads Engelhardt as defending the principle of permission as being the minimal content necessary for morality, one must ask, "Why this content and not some other?" One could reply that this is the minimum assumption necessary to resolve moral conflicts with secular moral authority, which is essentially what he argues when he argues that the principle of permission is a necessary condition for secular moral debate. If true, this would be a powerful reason to hold that this assumption, and no other, ought to be the starting point for secular moral discussion. However, it is not true.

The principle of permission is not the only method that can maintain some form of conflict resolution in a pluralistic society that has a "family resemblance" to moral theories. Granted, it is necessary to have something like the principle of permission—really, it is necessary to have more—in order to make secular moral arguments be much use in resolving difficulties, but Engelhardt rejects importing content merely to make a secular moral theory more useful. What is important is that the principles of the secular bioethics be content-free—which, on the interpretation being pursued here, must mean, "as content-free as possible." In fact, a principle significantly less contentful than the principle of permission can make secular moral discourse possible, while still retaining that family resemblance to morality, although it cannot make much *peaceful* secular moral discourse possible. The principle of permission is thus not necessary for secular moral discourse.

---

[103] Wildes, "Engelhardt's Communitarian Ethics: The Hidden Assumptions", pp. 77–93; Aulisio, "Contingency and Relevance."

[104] See Engelhardt, "The Foundations of Bioethics and Secular Humanism: Why is there No Canonical Moral Content?" p. 265, where he notes that "One cannot require permission to carry with it any particular content or will to beneficence without undermining the project itself by begging the question."

## The Principle of Reason-Giving

One could plausibly maintain that morality is essentially a process of providing good reasons for one's claims and rationally defending one's actions. This requires being able to respond to questions about one's moral beliefs. If one presents a moral claim, it is a legitimate move for another moral agent to ask why one believes it. The reasonable move is then for the first agent to give a supporting reason that gives intuitive or deductive support for the first claim. This is consistent with the analysis that Engelhardt gives of the practice of secular morality as requiring justification to and for all persons. At first glance, it seems that all claims are subject to question, and thus that a minimal morality cannot contain any basic claims, be they the principle of permission or anything else, without making a contentful claim.

But one might try to defend a minimal theory of morality by making the following sort of argument: Not all beliefs are subject to such rational questioning. If the justification for a belief is, "Because holding this belief is necessary for the coherent continuation of the ability of persons to perform the activity of giving and asking for reasons," then one *cannot* coherently ask, "Why ought I to believe that?" The very act of posing the question presupposes the rational exercise of reason-giving; by posing the question and expecting a coherent answer one is necessarily assuming all that is needed for the exercise of rationality and reason-giving. Thus, any belief that is justified by the necessity of holding that belief for the continuation of rational discussion cannot be rationally questioned. If the practice of coherent reason-giving and asking is rejected, then answers need not relate to the questions, and the very practice of morality (so conceived) will cease to be possible; thus, this practice is the fundamental, necessary part of such a practice.[105]

Rationality, in the form of giving and asking for justifications of beliefs and actions, then, can provide a kind of foundation, or basic grammar, for the practice of ethics. If one participates in such a practice, then one participates in giving and asking for justifications, though the reasons given may not be convincing for all persons and though many beliefs may be founded on assumptions that are not, themselves, justifiable to the others in a discussion. From this participation, one can then conclude that there can be a basic principle that cannot be coherently questioned. Call it the "principle of reason-giving," which would be defined as something like, "One ought to be able to provide a coherent reason for one's actions, and coherent reasons for one's justifications. If one cannot, one cannot morally perform that action." Precisely what this principle would be, and what it would entail, would take some working out which, for reasons made clear in *Back to Nihilism?* below, is not attempted here. All that is shown here is that (a) this is a potential fundamental

---

[105] One may, of course, choose to reject altogether the concept of giving and asking for reasons. Nothing in particular requires that one be rational. But the practice of morality in a secular pluralist environment, where one is likely to be frequently challenged and must be able to defend one's claims, requires that one be rational in at least this minimal sense. All this shows is that reason-giving is necessary for the practice of secular morality, not that it is somehow necessary in and of itself.

grounding for secular moral discourse, though not a very effective one, (b) it shares a family resemblance with traditional morality, and (c) it is not the same as, and less contentful than, the principle of permission. If all of these things are shown, then it will have been shown that the principle of permission cannot be the basic minimum necessary for secular moral debate.

One could appropriately defend, on grounds of the principle of reason-giving, the principle that force provides no intellectual justification. When someone asks for a reason, whacking him with a baseball bat is not an acceptable response. Still, though it is possible to derive from the principle of reason-giving the conclusion that "Force cannot prove claim X to be morally correct," one cannot thereby show that force cannot be used unless justified to all persons involved—e.g., that one may not use force against unconsenting innocent others. If so, the principle of reason-giving is not the same as the principle of permission.

The Argument for Force shows the limits of the principle of reason-giving. There is nothing wrong in the structure of this argument, nor in the reason-asking and giving that a proponent of such an argument could engage in. Admittedly, the proponent of this argument would reach a conclusion with which many others would not agree, that violence against a doctor is acceptable if that prevents some abortions from happening; but his line of argument is no different, on the grounds of reason-giving, than any other chain of justifications that does not conclude in some version of, "Because this is required for the practice of rational reason-giving." Neither allowing abortions nor prohibiting them, even by means of force, can be restricted by the practice of reason-giving; both are among the many things completely consistent with that practice.

The principle of reason-giving shares the family resemblance with traditional moral theory in much the same way as does the principle of permission. It addresses traditionally moral issues such as whether one may or must kill another, save another, and the like. One must perform these acts when it is necessary for the practice of reason-giving, must not when it is contrary to that practice, and may or may not when either could be supported by reason-giving. Admittedly, many of the traditional moral issues will fall into the latter camp, where one either may or may not perform them, but the same is true of the principle of permission. For both principles, rules explicitly forbidding many practices cannot be defended. The principle of reason-giving also allows for praise and blame. One is to be praised for having valid justificatory reasons for one's actions, and blamed for failing to have them. Finally, it shares the logical progression from one claim to another with traditional theories: basic claims required by the practice of reason-giving justify some other claims. Other basic (but contentful) moral claims can also be shown to justify other claims by providing good reasons to believe them. The principle of permission shares this family resemblance similarly, but no more so. It addresses moral issues, provides for praise and blame, such as when it is right and wrong to employ force, and allows further claims such as the theory of ownership to be justified if they proceed from the principle of permission. The principle of reason-giving seems to share as much of a family resemblance with morality as does the principle of permission.

It can also be shown more directly and more generally that using force against unconsenting others cannot be rejected on the grounds of the principle of reason-giving. It is consistent with reason-giving to hold that a set of beliefs is correct, and that they are correct for all persons, and possibly in all times. But it does not follow from this that reasoned discussion is the only rationally acceptable means to use in the pursuit of spreading one's moral beliefs. Consider, for example, the deceit and trickery that such utilitarians as Sidgwick and Parfit suggest may be necessary to truly pursue the most good for the most people.[106] Nothing in the notion of giving and asking for reasons prohibits this, once one has concluded that the highest moral good is in the promotion of the greatest good for the greatest number. Indeed, one can give reasons for doing such a thing: this deception promotes the most good for the most people. Force, or at least trickery and deceit, would be justified by reason-giving.

More generally, the following seems like the most plausible argument for preferring agreement over aggression in a pluralistic society:

(1) In a pluralistic society, it is understood that not everyone will share the same moral beliefs, nor act the same way. Even if one attempts to convert others to one's view strenuously, this will still be true. (Assumed as a hypothesis)

(2) Given (1), one must either accept that other persons may at times act against one's own personal moral standards (tolerance), or one may refuse to accept it.

(3) If one refuses to accept this fact, one may either attempt to change this or be eternally frustrated.

(4) It is better to remain tolerant than to employ forceful actions to make others' actions conform to one's moral standards.

(5) It is also better to remain tolerant than to be eternally frustrated.[107]

(6) Consequently, one ought to be tolerant of other moral standards.

One could, of course, support this view by giving reasons for why it is best to follow this argument. But one could also give reasons for why it would be best to follow this argument:

(1*) In a pluralistic society, it is understood that not everyone will share the same moral beliefs, nor act the same way. Even if one attempts to convert others to one's view strenuously, this will still be true. (Assumed as a hypothesis)

---

[106] Sidgwick's *Methods of Ethics* and Parfit's *Reasons and Persons* both explore the possibility of deception as a means to furthering the greatest good for the greatest number. See Parfit, Derek. (1984). *Reasons and Persons*. Oxford: Clarendon Press, pp. 41–43 and Sidgwick, Henry. (1981). *The Methods of Ethics*, 7th ed. (1907). Indianapolis: Hackett Publishing Co., pp. 489–492.

[107] Although it is essentially grudging tolerance to refuse to accept the facts but not to attempt to change them by all means possible, this option is included because it seems relevantly different from accepting tolerance as a good thing in itself.

(2*)     Given (1), one must either accept that other persons may at times act against one's own personal moral standards (tolerance), or one may refuse to accept it.

(3*)     If one refuses to accept this fact, one may either attempt to change this or be eternally frustrated.

(4*)     It is better to employ forceful actions to make others' actions conform to one's moral standards than to be tolerant.

(5*)     It is also better to employ forceful actions than to be eternally frustrated.

(6*)     Consequently, one ought to employ forceful actions to change the standards and behavior of others.

The principle of reason-giving, therefore, cannot be used to defend the principle of permission. Yet it is a basis for a system of secular moral claims, though a very limited set, and grounds for praise and blame. So one must return to the question that must be asked of Engelhardt: why ought one prefer the principle of permission to some other principle or set of principles when seeking secular moral resolutions to moral conflicts? It cannot be because the principle of permission is the basic minimum principle necessary for secular moral reasoning—the principle of reason-giving is more minimal, yet is capable of grounding secular moral reasoning. It cannot be that the principle of permission is the only principle of secular conflict resolution that shares the family resemblance of morality—the principle of reason-giving shares that, too. And it cannot be because the principle of permission is a principle that allows resolution of all or most secular moral difficulties, because it does not succeed in doing that. There is not, in short, a good reason to prefer the principle of permission as a means of secular moral conflict resolution. It is too contentful to be a purely procedural principle, and not contentful enough to be an adequately useful principle.

## Back to Nihilism?

If the principle of permission is truly a contentful principle, then content-free secular ethics is left with really only one option. One could accept the principle of reason-giving as the truly content-free procedural principle and allow only that principle to be employed in negotiations with moral strangers. The only other option that follows from the above is to surrender the concept of a content-free ethics entirely.

If one continues to seek a theory built from content-free principles, the results will be disappointing. The resultant secular moral theory will not be much better than the nihilism in which Engelhardt suggests, but for the principle of permission, the failure of the enlightenment project leaves us. The principle of reason-giving is a terribly unhelpful principle of morality. Indeed, for producing resolutions to morally problematic cases, it is inferior even to the principle of permission. It will not prevent any consistently argued yet atrocious claim from being "morally" acted upon. Yet, as Engelhardt himself notes, what he seeks is not a useful or preferable moral theory, but rather only the categorical possibility of a secular moral theory. If a moral theory is made possible by the principle of reason-giving, that it is a moral theory to which few would have any motivation to appeal is not relevant.[108] But it is not the helpfulness of the principle of reason-giving that is the rationale for its being accepted. It is a content-free, procedural principle. The point that is relevant is that it has the same strength of justification as Engelhardt argues the principle of permission has, and that I have argued above the principle of permission does not actually have. Though there are serious problems with both systems, he cannot argue that the principle of permission is the only possible means to resolution of problems that can appeal to a secular moral authority. In fact, the principle of permission has less claim to that, as a content-thin principle, than does the principle of reason-giving. But if the principle of reason-giving is what secular moral debate has to which to appeal, it has very little of use indeed.

The only answer Engelhardt can give to prefer the principle of permission seems to be that the principle of permission is the principle required to address secular moral questions *peacefully*. But once one accepts this, one must then ask why this is what one ought to do. Peaceful resolution is what most persons aim for most of the time, but why ought one to accept a peaceful but obviously (to one's own eyes) wrong resolution rather than a non-peaceful, "correct" one? The only reason is that there is some value to peaceful resolution in and of itself. To accept this is to embrace the other option noted above, that of rejecting content-free principles in favor of a "content-thin" one. Engelhardt must make this minimally contentful claim at the base of his theory—a minimal, reasonable contentful claim to which many would easily accede, but a contentful claim nonetheless. It is a way out of nihilism, and a way out of the nearly-nihilism that the principle of reason-giving leaves us with, but it requires content.

Rejecting the hopes of a secular ethic guided by content-free principles is perhaps not such a bad thing. If the secular morality that could be defended by the principle of reason-giving alone were the only possible secular pluralistic means of resolving morally problematic cases, secular medical ethics would be in desperate straits indeed. As Engelhardt shows, secular morality may not need to be so limited. But in order to avoid these limitations, basic, minimally contentful assumptions must be made; if one chooses to do this, one ought to choose one's content-thin principles well. Since the principle of permission is not, therefore, the only possible means of attaining moral authority in a secular, pluralistic society, the fact that it is, first, a substantial moral claim and, second, is incapable of resolving many problems, gives good reason to look elsewhere for a means of addressing morally problematic cases in modern society.

---

[108] See comparable commentary with regard to the principle of permission in Engelhardt, "The Foundations of Bioethics and Secular Humanism: Why is there No Canonical Moral Content?", pp. 264–265.

# Chapter 3
# The Four-Principles Approach: An Appeal to the Common Morality for Resolution and Justification

As explored in Chapter 2, the minimal requirements of content-free secular morality are too minimal to serve as any kind of useful tool for justifiably solving moral conflicts in a pluralistic society. Tom Beauchamp and James Childress try a different approach to address the same problem. They argue that what they understand as the "common morality," held by all morally serious persons, already contains certain basic norms, which can be loosely codified as principles of morality. These principles, appropriately made specific and in interaction with each other, can be used to respond to and resolve moral conflicts.

However, it is argued below that the use of principles taken from this "common morality" to resolve moral problems is not sufficient in a pluralistic society. The use of principles to determine a correct action or to justify that action in a particular case requires the use of prior moral beliefs that are not present in a commonly shared common morality, and so are not shared by all reasonable persons in a pluralistic society. Judgments made by the use of principles cannot be justified to persons who do not share the same prior beliefs; when principles are used to determine the right action in a case they cannot be successfully used to justify that action in a pluralistic context.

This conclusion does not mean that no interesting conclusions can be derived from the use of principles, nor does it mean that principles cannot be used in an important way in a pluralistic society. Principles can be useful in explaining, elaborating, and resolving difficult moral problems, though only in the context of some shared basic moral beliefs. But the means that Beauchamp and Childress use to justify these solutions—reflective equilibrium between initial considered judgments and later specified systems[1]—depends upon a considered judgment about the principles in at least a general form, an agreement which does not exist across a pluralistic society. The conclusion to be drawn from this argument is not that one can never employ principles usefully in a pluralistic society, but only that the justifica-

---

[1] Beauchamp, Tom L., and James F Childress. (2001). *The Principles of Biomedical Ethics*, 5th ed. New York: Oxford University Press. [hereafter, Beauchamp and Childress, *Principles*, 5th ed.], pp. 20–26. Unless otherwise noted, all references here are to the fifth edition. See further discussion of Beauchamp and Childress's justification below.

S.S. Hanson, *Moral Acquaintances and Moral Decisions*, Philosophy and Medicine 103, DOI 10.1007/978-90-481-2508-1_3, © Springer Science+Business Media B.V. 2009

tion for the conclusions derived by them, and the actions proscribed by principle-based arguments, cannot be justified by an appeal to a coherent common morality system of specified principles. These decisions thus will not necessarily be adequate to resolve issues with moral authority between moral strangers. This does not prevent principles from being useful in the context of first-person justification or justification within moral communities and other groupings as well; principles can be a means of addressing difficult moral cases in these circumstances.[2] Because use of principles in this context may not produce resolutions that will be understood to be correct by all reasonable persons in a pluralistic society, this will not make principles able to resolve morally problematic cases for all in a pluralistic environment; but it can provide a means for at least some persons and groups of persons to so use principles successfully to resolve morally problematic cases.

Tom Beauchamp and James Childress argue that the four principles of respect for autonomy, beneficence, nonmaleficence, and justice, if appropriately analyzed and understood, can be used to address moral conflicts between persons who rationally disagree on moral theory and fundamental moral claims.[3] Beauchamp and Childress do not argue the implausible claim that all reasonable persons could appeal to a set of principles to resolve all morally problematic cases. They are explicitly aware that moral dilemmas will continue to exist and remain unresolved after careful reflection.[4] Their goal is to produce a "method that helps in a circumstance of conflict and disagreement."[5]

More precisely, Beauchamp and Childress argue that a pluralistic principle-based account can and should be derived from what they call the "common morality"—a concept which will need some examination—and adjusted under of a model of reflective equilibrium.[6] The principles are derived from the common morality and are utilized to address specific moral cases or problems; the principles are then balanced and/or specified with regard to these particular cases or problems, and the procedure is repeated as new cases or problems arise. The principles develop and change, at least in terms of their specificity, and are always understood as both *prima facie* binding and subject to revision as needed.[7] When a particular case does not involve a specific conflict between the principles as they are currently specified and

---

[2] This is discussed below in Case 3.2, and in more depth in Chapter 5.

[3] Argued in Beauchamp and Childress, *Principles*, 5th ed. The changes made in the 6th edition do not significantly change this part of their theory.

[4] Beauchamp and Childress, *Principles*, 5th ed., p. 11; Beauchamp, Tom L. (2003b). "A Defense of the Common Morality." *Kennedy Institute of Ethics Journal* 13(3): 259–274. p. 268.

[5] Ibid.

[6] Beauchamp and Childress, *Principles*, 5th ed., pp. 398–99, 401–406. Reflective equilibrium, as a concept, is derived by Rawls, John. (1971). *A Theory of Justice*. Cambridge, MA: Harvard University Press, p. 49. The term "wide reflective equilibrium" was coined to describe this mode of reasoning in Rawls, John. (1975). "The Independence of Moral Theory", Presidential Address, *Proceedings and Addresses of the American Philosophical Association* 48s: 8.

[7] Beauchamp and Childress, *Principles*, 5th ed., pp. 14–16.

the principles as specified defend a particular resolution to the case, they may be utilized to resolve the case morally; when the case does involve such a conflict, or the action required is not clear, the principles are to be further specified or balanced to address and resolve the conflict.

One aim of this reflection is to find the most coherent system of moral rules, principles, and judgments by adjusting and adapting the system as new cases and reflections arise. The principles initially fit into this system as considered judgments which serve as beginning points for reflection; they are specified and developed into more appropriate and complex norms to augment the more general principles.[8] Another clear aim of this process is to develop a means of providing morally justified resolutions to particular cases; and the practical and the theoretical hands work together. Each new case brings a new moral judgment that needs to be fitted into the coherent whole; but the base of moral rules, principles and judgments in the coherent system also provides a means to resolve those new cases and to justify that solution.

The move to discussion of moral problems via common-morality-based principles that they believe should be shared even by Engelhardtian "moral strangers" functions as an attempt to address the issue of resolving moral conflicts in a society where persons do not necessarily share similar moral intuitions, theories, or case judgments. Though different persons may view the moral landscape differently through their particular moralities, in the shared common morality certain principles are agreed upon. These principles are less abstract than theory, but more general than particular case judgments or rules of action. By deriving a system for addressing moral problems based only on those shared "mid-level" principles, which are embraced in the common morality, Beauchamp and Childress attempt to provide a method that is calculated to facilitate resolution of moral problems[9] by reference to commonly accepted principles understood to be morally important by all morally serious persons.

If they are successful in justifying a system of problem resolution that retains all of the pluralistic appeal that it intuitively has at the beginning, then they will have resolved the motivating problem of this work, and the best means for initiating attempts to resolve moral conflicts between moral strangers will be the better understanding and development of this common morality theory. However, in many cases, their resultant system cannot resolve moral problems by both being able to determine a right action to take in a potentially morally difficult circumstance, and being able to justify that decision as correct to others in a way that those others can recognize as a valid justification. That is, they cannot resolve moral controversies.

Their approach has two options, neither of which will result in a means for all morally serious persons with rationally differing moral views to resolve moral

---

[8] Beauchamp and Childress, *Principles*, 5th ed., pp. 15–17. See also pp. 405–406.

[9] More precisely, they make only the claim that this account is intended to help resolve bioethical issues, and they justify this with various reasons why these principles in particular are well suited to bioethical resolutions. See the discussion along these lines in the following section.

conflicts with moral justification: first, it may develop into a clear set of carefully specified rules for action by thorough analysis of cases, rules, principles, and judgments and appropriate modification of the resulting set of beliefs, principles, and moral claims into a coherent, reflective equilibrium; or it may maintain its ability to appeal equally to persons with widely differing rational moral views. I argue that it is not able to do both for all or even most persons in a pluralistic society. When there is rational moral disagreement about the issues involved in a case, the four-principles approach may not always be a means for persons in a modern, liberal, pluralistic society to resolve troubling moral conflicts. Principles may be specific, or widely shared, but generally not both. Because the capacity to determine the correct action in given cases requires specific principles, and the grounds for justification of those actions to others requires that they be shared generally in the common morality, Beauchamp and Childress need the principles to be both shared and specific.[10]

Beauchamp and Childress ground the justification of principle-based resolutions of cases in the coherent system of principles. The common morality consensus on the principles allows them to serve as considered judgments with which to ground a coherence-based justification of actions to be taken in particular morally problematic situations. The justification of this common morality and principle-based theory depends on the coherence of the system and its source in the shared appeal to the principles in the common morality.[11] Their argument for grounding the justification for a principle-based system in the coherence of the overall system is convincing, but the general versions of the principles in the common morality are not a good source for providing a coherent system of principles that can justifiably resolve problems to all or most reasonable moral persons. Between moral strangers, there is not enough shared in the common morality to ground a coherent, shared set of principles; between moral friends, there is no need to build a coherent system based on the common morality when one can build a system based on the fuller amount of moral claims shared by those moral friends. Principles can often be used within the context of moral agreement to address morally problematic cases or problems, but without that context the justification for any resolution is lacking. Even in relatively trivial, non-dilemmatic cases, one cannot consistently appeal in a pluralistic context to principles that are both specified and commonly shared, because specification augments and changes the principles from their generally shared predecessors.

---

[10] Insofar as the principles are adequately specific and adequately shared to resolve a particular case with regard to the persons involved in the case, they can resolve that case. This is an example of a case resolution in the context of a moral acquaintanceship, which is discussed in Chapter 5. This can function in the context of specific moral acquaintanceships, but not in the broader sense wherein it encompasses all or most of a pluralistic society. See Chapter 5, *Principle-based Resolutions in Pluralistic Settings* and *Moral Community, Moral Friendship, and Moral Acquaintanceship*.

[11] See, e.g., Beauchamp and Childress, *Principles*, 5th ed., pp. 398–401.

## Why These Principles?

Since Beauchamp and Childress present the four-principles approach as "link[ing] a coherence theory of justification to a common-morality theory," it is appropriate to examine both of these concepts briefly to appreciate how the method is developed and is meant to function.[12]

Beauchamp and Childress derive their four principles from the "common morality," which is understood to be a "set of norms that all morally serious persons share" and "accept as authoritative."[13] Moral norms that are accessible to and agreed upon by all morally serious persons can serve as a means of grounding moral claims with moral authority for persons with differing backgrounds or moral views. This is a large part of the reason that Beauchamp and Childress utilize the common morality as the beginning point for moral reflection:

> The general norms ... found in philosophical ethical theories are invariably more contestable than the norms in the common morality. . . . Far more social consensus exists about principles and rules drawn from the common morality (e.g., our four principles) than about theories.[14]

If the common morality is indeed commonly shared, it could plausibly be a means for persons in a pluralistic society to address, and sometimes resolve, moral conflicts.[15]

But why are these four principles the ones derived from the common morality, which may well contain other principles as well? This decision was made with a practical goal in mind. Beauchamp and Childress are interested not in providing a complete theory of morality, nor a complete set of principles for all of morality; rather, they have a more focused aim. They seek to provide a method for moral thinking, including an approach to moral conflict resolution, in biomedical ethics. The four principles, which are better understood as "clusters" of principles or norms, along with some associated rules, are those that they have determined each play a vital role in health care ethics.[16] Their concern is to derive a resultant theory or system as a method of doing health care ethics.

Whether or not they seek to create a moral theory or the less-all-encompassing task of creating a bioethical theory only, it seems at least initially plausible that respect for autonomy, beneficence, nonmaleficence, and justice are all principles held to be significantly important in the common moral view, even in the face of

---

[12] Beauchamp and Childress, *Principles*, 5th ed., p. 407.

[13] Beauchamp and Childress, *Principles*, 5th ed., p. 3.

[14] Beauchamp and Childress, *Principles*, 5th ed., p. 404.

[15] Some doubts have been raised as to whether there is such commonality among persons; see, e.g., Engelhardt, *Foundations*, 2nd ed., pp. 56–57, where he suggests that much of the reason that common agreements can be derived from persons with "radically different" moral theories (e.g., teleologists and deontologists) is that, though their theories differ, their actual moral "lifeworlds" are similar. When there is significant difference in moral belief, he argues, these agreements are far less plausible. See further discussion in *Differing Specifications in a Pluralistic Society* below.

[16] Beauchamp and Childress, *Principles*, 5th ed., pp. 12–14.

the sort of pluralism discussed in the previous chapters. It may not yet be clear what import the common morality has, but the presence of these four principles in it seems likely, though at what level of abstraction and/or utility is not yet clear. Second, the moral concepts embodied by these principles (at least at some general and abstract level) are fundamental to biomedical ethics. The practice of medicine has for thousands of years been a practice of nonmaleficence and beneficence; if it were not important prior to this, the relatively recent development of the notion of informed consent has made it certain that respect for autonomy must also be important. As more and more attention is focused on minority rights, women's rights, proper care of various populations of marginalized power such as the homeless or prisoners, and fairness in distribution of health care resources and information, a concern for justice too must play a major role in health care ethics; Beauchamp suggests that perhaps it must now play much more of a role than it has played before.[17]

There has been little challenge to whether the principles, in some form, are moral principles, and no challenge will be provided here either. A more troubling question for the four-principles approach is whether, because of the level of abstraction required to claim that the principles are widely shared, the four-principles approach can achieve anything like the goals of finding a justified method that can often help resolve bioethical problems in the face of cultural moral pluralism and pluralism about moral theory.

## Two Versions of the Common Morality

The first concern with the four-principles approach is their source. As noted, Beauchamp and Childress argue that their four principles are ultimately derived from what they call the "common morality." This concept can be interpreted in at least two ways, which will have an important impact on the method of moral resolution that the principles can provide.

David DeGrazia has distinguished between what he calls "common morality 1," which is the set of moral beliefs that are in fact widely shared by all morally serious persons, and "common morality 2," which he describes as:

> the set of moral beliefs that probably *would* be widely shared among morally serious people who give initial credence to considered judgments … and expand their moral thinking in conformity with the criteria for evaluating moral theories—e.g., consistency and argumentative support.[18]

Common morality 1 is a descriptive set of ethical values and ideals, and as such might be quite narrow, perhaps, DeGrazia argues, containing only

---

[17] See Beauchamp, Tom L. (1994). "The Four-Principles Approach," in *Principles of Health Care Ethics*, Raanan Gillon, ed. New York: John Wiley & Sons, pp. 3–12.

[18] DeGrazia, David. (2003). "Common Morality, Coherence, and the Principles of Biomedical Ethics." *Kennedy Institute of Ethics Journal* 13(3):219–230. pp. 221–222. Italics in original.

nonmaleficence.[19] Common morality 2 might be more expansive, as the fact that some persons would disagree with some of its content would not intrinsically render that content suspect. Beauchamp and Childress are at times unclear as to whether they appeal to common morality 1 or common morality 2 as the source of their principles, and may in fact muddy the distinctions between the two concepts.[20]

Beauchamp argues that the principles found in the common morality are justified because:

> ... they are the norms best suited to achieve the objectives of morality. Ultimate moral norms require for their justification that one states the objective of the institution of morality. Once the objective has been identified, a set of standards is justified if and only if it is better for reaching the objective than any alternative set of standards.[21]

If one interprets the objective of morality or of principle-based theories as being able to resolve moral difficulties in a pluralistic society, neither a principle-based theory grounded in common morality 1 nor one grounded in common morality 2 can truly claim to be "better for reaching [that] objective than any alternative." It may be the case that this is not the goal the Beauchamp and Childress have in mind, but since this is the goal of this work, it is important to examine whether principles can serve that purpose better than any alternative.

Common morality 1 is a mere description of the widely held beliefs that exist in the minds of persons who care about acting morally, whom Beauchamp and Childress refer to as "morally serious" persons. This does not, by itself, indicate that a principle-based approach that begins with a derivation of the principles from this set of beliefs is relegated to a merely descriptive and not proscriptive role[22]; principles are meant to be developed, specified, and coherently balanced in such a way that they provide ethical proscriptions that differ, perhaps greatly, from beliefs initially held either by individuals or the general populace. Were Beauchamp and Childress's derivation of basic principles to be shown to come from a descriptive common morality 1, that would not be a damning indictment of their system, as the principles after lengthy specification, etc., are not merely descriptive. Beauchamp and Childress need not imply that agreement on the principles entails justification— DeGrazia reminds us that even if 99% of the population share an erroneous moral belief, it is still wrong[23]—but rather that agreement on the principles makes them a good beginning point to the conversation.

In addition, by restricting the discussion to "morally serious" persons one may perhaps assume that these persons have reasons for holding these moral claims that

---

[19] Ibid., p. 221.

[20] Ibid., pp. 221–225.

[21] Beauchamp, Tom L. "A Defense of the Common Morality." p. 266.

[22] This conclusion is contrary to DeGrazia's claims about the role of a morality built upon common morality 1. He emphatically states that a reliance on common morality 1 reduces normative ethics to descriptive ethics: see "Common Morality, Coherence, and the Principles of Biomedical Ethics." p. 224.

[23] DeGrazia, "Common Morality, Coherence, and the Principles of Biomedical Ethics." p. 222.

are consistent with good moral thinking. Perhaps one need not depend, in other words, on the mere agreement of these persons but on the seriousness of the thought that has gone into their defense of the agreed upon principles. If so, then common morality 1 may serve as a legitimate basis on which to build a principle-based approach to morality.

Beauchamp and Childress may not be doing so, however. There is some evidence that Beauchamp and Childress employ common morality 2—e.g., Beauchamp states, "For purposes of empirical investigation, my claim is that all persons committed to morality, and all well-functioning societies, adhere to the general standards of action enumerated previously."[24] The qualifications of "commitment to morality" and "well-functioning societies" may be thought to limit this to those persons and societies that have met the criteria for common morality 2. Yet, the evidence that they employ version 2 may be unclear, as DeGrazia argues.[25] There is a third possibility, which is that Beauchamp and Childress may be working with a form of the common morality that employs both of DeGrazia's common moralities 1 and 2. Recognizing that a development of a coherent theory from abstract beginnings through the lengthy process of balancing and specification is an ongoing, and never-ending, process,[26] Beauchamp and Childress may be appealing to a common morality that has aspects of version 1 in using the principles as provisional starting points, but aspects of version 2 as the principles are developed and specified. This would render consistent the perceived inconsistencies in their use of the phrase, as well as be more consistent with their own view of the development of moral theory in general and principles in specific.[27]

In any case, though some different concerns will apply to Beauchamp and Childress's theory depending upon which version of common morality they appeal to, each version will have similar problems for use as a system to resolve moral conflicts in a pluralistic society. These concerns will be laid out here and argued for more carefully in the following sections.

The principles that one might derive from a common morality 1 agreement will need quite a lot of specification. Beauchamp's own interpretation of the content of the common morality is that it "contains only general moral standards that are conspicuously abstract, universal, and content-thin."[28] The principles that could be derived from common morality 1 will indeed be exceedingly vague and initially unhelpful in addressing moral problems. It may be the case that all morally serious persons agree that, for example, beneficence is a moral principle. Still, the principle so named will differ radically between morally serious persons. As noted in Chapter

---

[24] Beauchamp, "A Defense of the Common Morality." p. 262.

[25] DeGrazia, "Common Morality, Coherence, and the Principles of Biomedical Ethics." p. 222.

[26] See, e.g., Beauchamp and Childress, *Principles*, 5th ed., p. 398: "We can never assume a completely stable equilibrium. The pruning and adjusting occur continually in view of the perpetual goal of reflective equilibrium."

[27] Beauchamp and Childress, *Principles*, 5th ed., pp. 404–405.

[28] Beauchamp, "A Defense of the Common Morality." p. 260.

2, the principle of beneficence accepted by an Engelhardtian libertarian will differ dramatically from that accepted by a Singerian utilitarian. Each is morally serious, and each accepts a principle of beneficence, but their interpretation of the meaning of that principle, and any specifications or weightings of it, will be quite different. The same will be true for other principles. In a vague and unhelpful form they may be held by all morally serious persons; but they are not so jointly held in a uniform, contentful, and useful form.

Moreover, the balancings and specifications that will seem appropriate to those different serious moral thinkers will tend to make the principles diverge rather than converge. What might have begun as universally shared will rapidly become unshared, not because some persons would refuse to engage in the thoughtful specification of the principles but precisely because they would so engage.

This would reintroduce the problem of unshared moral content to a theory, at which point the justification for that content would have to be provided. The justification for preferring one form of specification over another is limited to improved coherence. In light of the wide-spanning yet coherent disagreement about the content of a moral life, further developed in arguments below in *Specification and (Lack of) Universal Agreement*, it is doubtful that one carefully specified and balanced set of principles could be shown to be the most coherent and consistent. A system of principles that depends on common morality 1 will devolve into multiple, incompatible sets of principles, none of which could lay claim to superior justificatory status.

Conversely, a set of principles based upon common morality 2 would not necessarily begin as widely shared, and would certainly not become widely morally shared once specified. But here, in opposition to the theories derived from common morality 1, one could argue for the superior justificatory status of one specified set of principles. When persons disagreed with a particular interpretation or specification of the principles, as would be sure to occur, one could argue that a particular set of beliefs is that which "*would* be widely shared among morally serious people..."[29] who consider the principles and specifications thereof properly. A set of principles derived from common morality 2 could retain justificatory status as the only *appropriately* specified and balanced set of principles. This resolves the ambivalence of the principle-based approach sourced in a common morality 1.

Unfortunately, it does so at a high price, for it introduces unshared moral content that is simply unjustifiable on shared moral grounds. Even morally serious persons, if they have widely divergent understandings of the moral world—which they do—are going to arrive at divergent specifications of the principles and of the decisions made by those principles. As shown below, especially in *Specification and (Lack of) Universal Agreement*, it is implausible to argue that all morally serious persons will or even could end up with only one specified set of the principles, if only they would think about it carefully and in conjunction with the rules of coherence discussed by Beauchamp and Childress—e.g., consistency, completeness, explanatory

---

[29] Ibid.

and justificatory power, etc.[30] What seems far more likely to occur is that persons with divergent worldviews will understand divergent coherent sets of the principles to be the lone set that would be accepted by all morally serious persons, which will not allow principles alone to serve as a method of bridging the moral gap between them. Not only will persons not end up sharing the same set of specified principles, but also, they could not do so. The same system would not cohere with different persons' divergent worldviews.

Beauchamp has the following to say about the difficulties of specifying principles in the context of conscientious disagreement:

> There is always the possibility of developing more than one line of specification when confronting practical problems and moral disagreements. Different persons and groups will offer conflicting specifications. In any given problematic or dilemmatic case, several competing specifications may be offered by reasonable and fair-minded parties, all of whom are committed to the common morality.... Conscientious and reasonable moral agents understandably will disagree with equally conscientious persons over moral weights and priorities in circumstances of a contingent conflict of norms, and many particular moralities therefore will be developed.[31]

Which is followed later by:

> Competing specifications generate moral disagreement and conflict. The moral life always will be plagued by forms of conflict and incoherence. The theorist's goal should be a method that helps in a circumstance of conflict and disagreement, not a panacea.[32]

However, it is hard to see how an appeal to the common morality provides a "method that helps" if it produces multiple conflicting sets of instructive specifications with no justified reason to prefer one over the others. And, as can be seen from the above, using a blend of common morality 1 and 2 would do little to alleviate these difficulties, as it is the process of specification and balancing that introduces and augments the likelihood of multiple coherent sets of principles.[33]

## Conflict of Principles, Balancing, and Specification

Since specification and balancing are seen to be problematic for the four-principles account, it is important to understand why they are also essential for the four-principles account to work. In virtually any case of moral significance more than

---

[30] Beauchamp and Childress, *Principles*, 5th ed., pp. 339–340.

[31] Beauchamp, "A Defense of the Common Morality." p. 268.

[32] Ibid.

[33] Beauchamp and Childress do note in a section labeled, "Problems for Common Morality Theory" that "attempts to bring the common morality into greater coherence through specification risk decreasing rather than increasing moral agreement in society." (p. 407) This is too weak a claim, as it is virtually certain that this shall happen, given the wide variation in worldviews in society; it is only in narrower circumstances that specification will help bring moral agreement. See fuller discussion below in *Specification and (Lack of) Universal Agreement*.

one of the principles will, at least *prima facie*, apply. Beauchamp and Childress give a good example of this on p. 342, summarized here:

**Case 3.1: A Refusal to Donate** A five-year-old girl has progressive renal failure, and her doctors determine that she needs a kidney transplant to give her the best chance at long-term survival. She is a difficult match; only her father is an acceptable, compatible related donor. The nephrologist discusses the matter with the father, noting that his daughter's prognosis, even with the transplant, is "quite uncertain." He chooses not to donate, and asks the doctor "to tell everyone else in the family that he is not histocompatible" because he fears that their knowing he refused to donate a compatible kidney would wreck the family.[34]

At least three, and probably all, of the principles come into play in the analysis of this case. The father is expressing an autonomous wish not to donate a kidney. The donation, if made, could help the child; though her prognosis with the transplant was not clear, it was also probably her only chance at long-term survival. The request that the doctor not tell the family the truth about the father's compatibility is phrased as a request that the doctor not cause harm, and thus falls under the broad category of nonmaleficence. A plausible argument could be made that the family has a right to be told the truth, even if this is painful, on grounds of respect for their autonomy. An appeal to criteria of justice may also come into play in determining whether the daughter is owed a kidney by her father, or whether it is fair to wait for a cadaveric kidney, which could otherwise go to another, when a compatible kidney is available here.

The point to note is that all of the principles are, in some form, relevant to this case, and it is therefore at least initially unclear what resolution of the case is correctly justified by those principles. Since Beauchamp and Childress are explicit in rejecting any hierarchical ranking of the principles,[35] one is left with the fact that there are three or four competing, possibly conflicting, moral principles in play in this difficult case, a problem which is shared by many other cases as well. How is one to determine how to use the principles in such a way as to make a justifiable resolution of this case?

The difficulty is in fact deeper. As Beauchamp notes with regard to a different issue:

[I]f nonmaleficence is the principle that we ought not inflict evil or harm, this principle does little to give specific guidance for the moral problem of whether active voluntary euthanasia can be morally justified. If we question whether physicians ought to be allowed to be the agents of euthanasia, we again get no real guidance.... In addition to abstract principles there must be mediating rules that translate an ethical theory into a practical strategy and set of meaningful guidelines for real-world problems.....[36]

---

[34] Beauchamp and Childress derive this case from Melvin D. Levine, Lee Scott, and William J. Curran. (1977). "Ethics Rounds in a Children's Medical Center: Evaluation of a Hospital-Based Program for Continuing Education in Medical Ethics," *Pediatrics* 60: 205. They employ this case as a test case to show how various moral theories (e.g., act utilitarianism, Kantian theory, ethics of care) address a case.

[35] Beauchamp and Childress, *Principles*, 5th ed., p. 405.

[36] Beauchamp, Tom L. "The Four-Principles Approach," p. 10.

In other words, even in a case where one need only consider one principle, the principle alone cannot determine what the resultant decision ought to be. How does one, in a case of active euthanasia, cause no harm? Is the death of the patient in such a case a harm? Is the *earlier* death of the patient by active means a harm? A greater or lesser harm? These questions and the like are all necessary in order to employ just one principle in such a case, but the principle itself gives no means of doing this in a justified way.

The answer that Beauchamp and Childress give to both of these problems is the practice of balancing and specification of principles, always with an eye towards the overall coherence of the entire system. Individual principles used in a vacuum are often not of much more use than admonitions to "Act virtuously" or to "Be good," though they are more directive. But when they are specified and interpreted in a system of principles, rules, and judgments they necessarily develop beyond their initial general level. By the process of specification, one takes the general principles and progressively delineates their content, deriving progressively more concrete rules in particular cases from them.

## Specification of Principles

Specification of norms is a method of interpreting them in particular cases, particularly cases of conflict between norms.[37] A specification of a norm or principle is defined as a focusing of the prior principle by means of "adding clauses indicating what, where, when, why, how, by what means, by whom or to whom the action is to be, is not to be, or may be done or the action is to be described, or the end is to be pursued or conceived"[38] in a way that helps to keep the central commitment of the original norm complete. The specification is a modification of the norm, but not, generally, a change in its fundamental meaning; rather, it is a modification to make it more specific, usually so as to make its original meaning more clear[39] and on point in a particular situation. The "commitments expressed in the initial norm" are to be expressed, if possible, in the specified norm; but the specified norm is more precisely articulated, particularly in response to a given case or a given conflict with another norm.[40] Specification does not guarantee continuity, but makes a strong effort to ensure it.

---

[37] Richardson, Henry. (1990). "Specifying Norms as a Way to Resolve Concrete Ethical Problems." *Philosophy and Public Affairs* 19: 279–310; Richardson, Henry. (2000). "Specifying, Balancing, and Interpreting Ethical Principles." *Journal of Medicine and Philosophy* 25: 285–307; DeGrazia, David. (1992). "Moving Forward in Bioethical Theory: Theories, Cases, and Specified Principlism." *Journal of Medicine and Philosophy* 17: 511–539.

[38] Richardson, "Specifying Norms as a Way to Resolve Concrete Ethical Problems", pp. 295–296.

[39] If the original meaning of the principle was overly vague in the first place, a specification can be more of a constructive effort to create a definite principle that reflects the basic moral notion that the original principle was meant to reflect.

[40] Richardson, "Specifying Norms as a Way to Resolve Concrete Ethical Problems", p. 297.

Specification is understood by its proponents to be an improvement in two ways over accounts of either deductive subsumption ("applying") of norms and/or intuitive balancing of norms. First, specification allows modification of norms as more cases involving the norm and potential conflicts are considered, which helps avoid problems involving inflexibility or historical stagnancy of norms; merely applying norms or balancing them without modifying them could retain an essentially vague or flawed principle.

Second, Richardson's account of specification also allows one to produce a more coherent system of principles and judgments. Since Richardson argues that bioethical theory must meet a minimal publicity requirement, meaning that reasons for one's specifications can be given, engaging in specification will produce a set of both specified principles and justifications of specifications.[41] This reasoning, along with the newly specified principles, produces a system of rules, principles, and norms along with the reasons that support them. This system can then be considered, as a whole, and brought into reflective equilibrium, making adjustments to the reasoning or the norms as necessary to achieve the most coherent rational whole. Specification, in addition to reflective equilibrium, can thus provide a potentially publicly available justification for a particular set of specifications and provide a means of justifiably rejecting other potential specifications. The more coherent a set of specified norms, the better justified that set is.

## The Objections of Clouser, Gert, and Green

When the four-principles approach includes specification and an appeal to justification by greatest coherence, it becomes a significantly different approach than it may have appeared before inclusion of specification. This allows it to respond to some objections; however, these objections reveal difficulties that will remain for the system to be employed in a secular, pluralistic society.

K. Danner Clouser, Bernard Gert, and Ronald Green have raised a serious set of objections to the four-principles approach, or to any similarly principle-based approach.[42] They provide multiple arguments that "principlism," as they call principle-based approaches, cannot function as a justifiable moral decision-making method, two of which are important here.

First, they argue that referring to principles to assist one in dealing with a moral crisis is practically useless. Principles "are primarily chapter headings for a discus-

---

[41] Richardson, "Specifying, Balancing, and Interpreting Ethical Principles", p. 285.

[42] The primary sources of this objection might be taken to be Clouser, K. Danner and Bernard Gert. (1990). "A Critique of Principlism." *Journal of Medicine and Philosophy* 15: 219–236; Green, Ronald. (1990). "Method in Bioethics: A Troubled Assessment." *Journal of Medicine and Philosophy* 15: 179–197; Clouser, K. Danner and Bernard Gert. (1994). "Morality vs. Principlism", in *Principles of Health Care Ethics*, Raanan Gillon, ed. New York: John Wiley and Sons, pp. 251–266; Clouser, K. Danner. (1995). "Common Morality as an Alternative to Principlism." *Kennedy Institute of Ethics Journal* 5(3): 219–236.

sion of some concepts which are often only superficially related to each other."[43] As such, they may be useful and often insightful as points to consider in moral reflection, but they are not well articulated for resolving problems and do not provide any guidance themselves on how to resolve any particular problem. Thus, they argue, any resolution of a moral question is dramatically underdetermined by the application of a principle, as it provides no direct guidance in particular cases; and the inclusion of another principle, rather than giving more guidance, gives instead more points to consider that are often contradictory to the points one considered regarding the initial principle. From this they conclude that whatever it is that finally guides one's decision in a case cannot be determined by the principles involved. Some other grounds, be it one's own moral views, hospital policy, fear of contradicting a colleague, etc., serve to decide the case. However, the principles are then appealed to as grounds for justification instead of what is actually used to resolve the case.

Second, they argue that since principle-based analyses provide no systematic method of justification, they yield "applications" of principles that are merely ad hoc. They argue that applying principles in an ad hoc fashion at best avoids utilizing any sort of justificatory theory and at worst obfuscates the need for and nature of such a theory and the practice of moral reasoning.[44] This, conjoined with what they call a "relativism" about the value of different historical moral theories that they see as inherent in the Beauchamp and Childress account, seems to encourage people to take whatever they see fit from various theories—e.g., the respect for autonomy prevalent in Kantian theories, the concern for the general utility from utilitarian theories, and so on—and ignore the lack of a coherent, justified system that this co-opting entails.[45] This concern generally seems to be Clouser, Gert, and Green's central concern with principle-based theories: that such theories deny an overarching theory to direct the employment of the principles, and thus remove any systematic justification of how principles are employed to resolve particular cases.

## Specification and Reflective Equilibrium as a Response

These objections seem to target an earlier version of Beauchamp and Childress's theory. Since Clouser, Gert, and Green's initial critique was made before the clear inclusion of specification and reflective equilibrium in the fourth edition of the *Principles*, it would seem that the criticisms would need to be rethought, as a significant change to the theory had been made. However, though they do begin to address specification, their critiques have remained essentially unchanged,[46] well after the pub-

---

[43] Clouser and Gert, "A Critique of Principlism", p. 221; see similar language in Clouser, "Common Morality", p. 223.

[44] Clouser and Gert, "A Critique of Principlism", pp. 228–230.

[45] Ibid., pp. 231–232.

[46] See Gert, Bernard, Charles Culver, and K. Danner Clouser. (1997). *Bioethics: A return to fundamentals*. New York: Oxford University Press; Gert, Bernard, Charles Culver, and K. Danner

lication of the fourth edition of the *Principles*, which clearly includes an important role for specification and coherence [47]—which remains largely unchanged in the fifth edition[48]—and also after DeGrazia's 1992 article on "specified principlism."[49] Their critiques also became more focused on Beauchamp and Childress's account of principles, which is reasonable considering the prevalence of that version, but amplifies the importance of addressing specification. Both of these objections have been significantly, and perhaps completely, addressed by the inclusion of specification of principles and reflective equilibrium; however, there are remnants of them that are still important for the project in this work.

The first objection, that "principles are chapter headings, but useless for addressing cases," is now essentially resolved. Though Beauchamp and Childress note that the principles begin at a reasonably vague and inchoate level, they are not left unchanged by the reflection that gets put into cases. Specification of a principle takes it from the vague level and progressively adds concrete specificity and analysis, and relates it to other principles; much work on this has been done already in the reflections in Beauchamp and Childress's chapters on each principle, as well as in much ethical reflection on cases elsewhere. Unspecified principles are something like chapter headings, though they are still *normative* chapter headings, but a specified[50] principle does not have that quality. (But see below in *A Further Difficulty* for a related problem that may still be present for specified principlism.)

The second objection is also mistaken, for a different reason. In their fourth edition, Beauchamp and Childress give a brief response to this concern, arguing that their intent was never to create a systematic theory, or to substitute for it, and plead the acknowledged fact that "even the core principles in [their] account are so scant that they cannot provide an adequate basis for deducing most of what we can justifiably claim to know in the moral life."[51] In the fifth edition, they expand on this to argue that any set of norms, including not only their own principles but also Clouser, Gert, and Green's moral rules, will "lack specific, directive moral substance" until properly specified.[52] They agree that a principle-based analysis "fails to provide an

---

Clouser. (2000). "Common Morality versus Specified Principlism: Reply to Richardson." *Journal of Medicine and Philosophy* 25(3): 308–322.

[47] Beauchamp, Tom L. and James F. Childress. (1994). *Principles of Biomedical Ethics*, 4th ed. Oxford University Press, pp. 23–32.

[48] Beauchamp and Childress, *Principles*, 5th ed., pp. 15–19 and 397–406.

[49] DeGrazia, David. (1992). "Moving Forward in Bioethical Theory: Theories, Cases, and Specified Principlism." *Journal of Medicine and Philosophy* 17: 511–539.

[50] The use of the term "specified" here does not imply that a single specification of a principle is all that is required. Specification is an ongoing, continual process, and a principle that has been specified before can and usually will be specified again later. See Beauchamp and Childress, *Principles*, 5th ed., p.17. I will use the adjective "specified" here to indicate merely that a given principle is one that has been specified and remains subject to potential further specification.

[51] Beauchamp, Tom L. and James F. Childress. (1994). *Principles of Biomedical Ethics*, 4th ed. Oxford University Press, pp. 106–107.

[52] Beauchamp and Childress, *Principles*, 5th ed., pp. 389, 404–405.

ethical theory, [but] see the criticism as correct but irrelevant,"[53] as they do not seek to do so and in fact are skeptical about any unified foundation for ethics.[54]

These responses may not resolve the central thrust of the objection, which is that without a single higher-level theory to unify and perhaps hierarchically structure the principles, many different and contradictory decisions can be made on the basis of principles without any justifiable means to choose between them. When, for example, a case of assisted suicide presents itself, the Clouser, Gert, and Green argument goes, no single answer is supported by careful thought about the various principles involved. One reflective individual might argue that it is right to provide a lethal drug to the patient because she requested it, and respecting her autonomy in this way is most important. Yet another might argue that it is wrong to provide it, because it could cause harm to her and her family, thus overriding her autonomous choice; a third might argue contrariwise that it is right to provide the lethal dose because it will be the most beneficent act towards her in the circumstance of her being terminally ill and requesting it. Each has analyzed the situation and presumably specified the principles involved, but arrived at a different answer. Moreover, this version of the objection holds, there is no guidance as to how to perform the specification or as to which, if any, of these accounts is correct, because such guidance could only come from a central unifying theory, which Beauchamp and Childress eschew.

It is precisely this limitation of specification to which Gert, Culver and Clouser point when they argue that "[s]pecification still fails to make critically important moves: listing and defending the morally relevant features of situations, [and] relating those features to a moral theory...."[55] Without being able to do the first of these tasks, and without being able to justify it on the grounds of the second, they argue, specification can occur for any reason whatsoever, whether moral or nonmoral. Racist, sexist, homophobic, etc. motivations are possible, and not in principle eliminable; principles can be specified in any way desired, with no grounds being possible for critiquing any person's specification.

This problem may be addressed, albeit not eliminated, by specifying principles not in a vacuum but in the context of seeking coherence in the system. In the four-principles approach, one is guided to make decisions and specify principles, not by a central unifying theory, but by the goal of seeking greatest coherence and reflective equilibrium. Clouser, Gert, and Green argue that a principle-based approach is essentially ad hoc because no unifying theory (such as a Gertian theory based on the principle of nonmaleficence) guides it; but the need for reflection and equilibrium makes the specification non-ad hoc. Beauchamp and Childress reject an overarching theory as a necessary tool for employment of principles (or other moral reasoning) not because they think no such theories are plausible or possible, nor necessarily

---

[53] Beauchamp and Childress, *Principles*, 5th ed., p. 390. Reference removed.

[54] Beauchamp and Childress, *Principles*, 5th ed., p. 390; see also Chapter 8, as a whole and esp. pp. 376–377.

[55] Gert, Bernard, Charles Culver, and K. Danner Clouser. (1997). *Bioethics: A Return to Fundamentals*. New York: Oxford University Press, p. 89.

because they believe all such theories are equally good.[56] Rather, they recognize that, on grounds of eight "conditions of adequacy for an ethical theory,"[57] various, though not all, different theories may be approximately equally rationally justifiable. The conclusion appropriately drawn from this is not that they therefore conclude that each is more or less equally right, but that each is plausibly well justified on the grounds that rationality alone can give for justifying moral theories.

Yet this does not mean that there is nothing which ties together the principles and their justification in a systematic way. Clouser, Gert, and Green are correct to argue that specification and balancing themselves do not do this, for one could, if one was merely considering the action of specification or balancing, specify or balance the principles in any one way rather than another. The act of specification itself places no restrictions on what specifications can be made. But specification in the context of seeking a reflective equilibrium among the resultant specified principles, rules, and judgments does restrict the choices one can make in a systematic way. One specification of principles may not be as good as another, because the resultant specification, case resolution, and related judgments will not fit as well into the coherent whole of the system. The justification of a particular specification, and the resultant resolution, then, is not that a particular case analysis is required by a unified overarching theory, but that it is required by reflective equilibrium. Whether or not one agrees with the means of justifying a moral system by appeal to reflective equilibrium, or whether one likes a system that is so defended, it is not possible to accuse specification of principles as having no means of systematizing the principles as they are specified.

Yet there is still some bite to this objection, if phrased another way. Users of the four-principles approach may specify in a non-ad hoc way by appealing to coherence and reflective equilibrium as a mode of comparison. But there is no guarantee inherent in the concept of reflective equilibrium that only one system can be rendered coherent. There remains the problem of multiple similarly coherent systems being possible.[58] In such a case, the choice between these systems might be seen as largely ad hoc. In truth, it would be far more likely that the choice between these two systems would depend on preconceived understandings of the world; but those notions are often presupposed and not rationally justified. Thus, if there were multi-

---

[56] They explicitly deny this in the fourth edition. Beauchamp, Tom L. and James F. Childress. (1994). *Principles of Biomedical Ethics*, 4th ed. Oxford University Press, p. 45. They are less explicit in the fifth edition, stating only that they do not agree to the hypothesis that all theories produce equally tenable moral frameworks. Beauchamp and Childress, *Principles*, 5th ed., p. 338.

[57] Beauchamp and Childress, *Principles*, 5th ed., p. 339–340.

[58] An additional concern exists. No system might achieve reflective equilibrium, thus requiring the process to continue forever without resolution. This is largely a theoretical concern, I believe, since one of the features to be considered in the reflection is how much time one has to reflect before making a decision. Equilibrium, especially in the way the Beauchamp and Childress employ the concept, is more of a comparative concept than a steady-state one, despite the implication of the system being "at rest" that the name implies. It is more of an ongoing comparative process where relative coherence is either improved or not by additions and alterations to a belief set. See Beauchamp and Childress, *Principles*, 5th ed., Chapter 9.

ple similarly coherent but contradictory systems, this challenge from Clouser, Gert, and Green could still be a concern. This is discussed further in Specification and (Lack of) Universal Agreement.

The principles, developed by specification and reflective equilibrium as suggested above, are normative in character, if not a moral theory per se. Resolving problems by consideration of the principles is thus not usurping moral reflection and replacing it with something else, but is an exercise in moral reflection. Reflection on the principles is reflection on moral concepts; moreover, Beauchamp and Childress argue, the four principles are all derivable from many legitimate moral theories. *Contra* the Clouser, Gert, and Green line that the four-principles approach, in affirming

> ...the principle of beneficence...acknowledges that Mill was right in being concerned with consequences[;]... the principle of justice ... acknowledges that Rawls was right in being concerned with the distribution of goods[;]... the principle of autonomy... acknowledges that Kant was right in emphasizing the importance of the individual person[; and] the principle of nonmaleficence acknowledges that Gert was right in emphasizing the importance of avoiding harming others...[59]

as if Mill and Kant never concerned themselves with matters of justice, etc., the derivation of each of the four principles comes from the common morality, which is common to holders of many different moral theories. Thus, a Millian utilitarian and a Kantian deontologist should both be able to recognize that each of the four principles is represented as something of moral value in the ethical theory that they hold. The principles are, then, a part of the content of morality, whatever theory it is that one holds.[60] The principles might be argued to function in the place of moral theory, especially where there is disagreement on moral theory, but they ought not be thought a non-moral replacement.

### Specification as a Loose Requirement

Richardson's definition of specification holds that one specifies moral norms or principles by "adding clauses indicating what, where, when, why, how, by what means, by whom or to whom the action is to be, is not to be, or may be done or the action is to be described, or the end is to be pursued or conceived."[61] But one might ask whether one must be so rigorous in order to use the principles to address at least some moral problems.

---

[59] Clouser and Gert, "A Critique of Principlism", pp. 232–233.

[60] Since Beauchamp and Childress no longer depend upon conceptual overlap in theories as a source of the principles, but rather on the common morality, this point is not emphasized in the fifth edition of the *Principles*. It remains as true as it ever was, though, even if not specifically addressed.

[61] Richardson, "Specifying Norms as a Way to Resolve Concrete Ethical Problems", pp. 295–296.

Consider, for example, the occurrence in the 1950s of physicians injecting cancer patients with non-therapeutic doses of plutonium.[62] In such a case, the principle of nonmaleficence is *prima facie* opposing the principle of beneficence: it was hoped that these injections would help educate scientists and physicians about the effects of radiation on the human body, which was thought importantly relevant at the time because of the cold war and the need to protect workers producing atomic weapons. One could address these cases by carefully spelling out, for example, a complex set of medical codes of behavior, including clauses that explicitly prohibit this sort of behavior by noting that nonmaleficence is more important than beneficence when non-therapeutic and dangerous experiments are contemplated, even when safety of others and national security is on the line, and even when the persons involved are thought to be terminally ill. Such a complex code would allow persons to resolve this case.

Perhaps, however, one does not need such a well-spelled out, complicated code. Perhaps one could simply argue that this is an unacceptable action on grounds of its obvious and blatant violation of the general principle of nonmaleficence, without the need for the intermediary of an explicit spelling out of the principles. Any decent person would know, from the principle of nonmaleficence itself, that such injections are wrong. It would take no careful, complicated spelling out to realize the correct response to this case. One might, the argument goes, simply deductively move from the principle of nonmaleficence to this conclusion; this conclusion follows directly from the principle itself. Consequently, one could resolve the issue of what to do in these cases, at least, to all morally serious persons.

This is not an attempt to defend deductive subsumption of the sort criticized by Gert, Green, and Clouser. Deductive subsumption directly from a general, vague principle cannot be a resolution of a problematic case, because it could not be justified, on the grounds of justification available to principles. Though it might be possible to derive this conclusion directly from the principle of nonmaleficence, one could not justify that derivation on grounds of an appeal to a coherent system, which is critical to the use of principles to resolve problems elaborated in this chapter. A pure deductive subsumption like this would be vulnerable to the critiques of Gert, Green, and Clouser, because that decision would not be grounded in a complex, coherent system. In any case, a pure deductive subsumption from the principle of nonmaleficence could easily be challenged by someone who deductively subsumes from the principle of beneficence that the defense of the country requires this sort of action. To choose one over the other by pure deductive subsumption would be at best dogmatic.

Rather, the point of the example seems not to be a defense of deductive subsumption, but a rejection of the need for explicit specification in the form of carefully crafted and added clauses appended to the principles in question. One might argue that all one needs in order to propose and defend the claim that one ought

---

[62] Described in Beauchamp, Tom L. (2001). *Philosophical Ethics*, 3rd ed. Boston, MA: McGraw-Hill, pp. 140–142.

not inject patients with non-therapeutic doses of plutonium is a more general recognition that, in cases like these, nonmaleficence is more important than the beneficence proposed. This need not be specified in any explicit fashion, by spelling out the various added clauses and modifications to the principles, but can be more generally recognized and balanced in cases like these without an explicit analysis. That is, one could argue that nonmaleficence is more important than beneficence in cases of non-therapeutic and dangerous experimental treatments. The principles of nonmaleficence and beneficence would remain unaltered, but their relevant importance in certain cases would be a part of the system that justifies the claim to be defended.[63]

If correct, this might mean that specification, per se, is not necessary in order to resolve a morally problematic case, though deductive subsumption cannot achieve this goal. A resolution must be justifiable, which for principle-based, common morality theories means it must derive from a coherent system, and some elaboration of the principles in order to make them more specific than their initial vague forms is necessary to make them into a coherent system. But perhaps this could come from some methods more general than the explicit form of specification, which requires modifying the principles themselves. Perhaps this could come from more generally weighing and balancing the relative importance of principles in certain cases like this one, without any explicit specification.

Whether or not such a system could be developed is unclear. What is important is that in what follows below, no assumption of explicit specification on Richardson's grounds should be inferred. A "coherent specification," as the term is used herein, means either that one has explicitly specified the principles in a coherent system, or coherently weighted or balanced them against each other in some other fashion, even if they have not explicitly specified them. If persons "share a coherent specification," they have a similar specific principle in a coherent system, whether that system was derived from explicit specification or otherwise. Similarly, "specifications" or "specified principles," or other variations on that theme, can refer to any coherent system of principles adequately specific to unambiguously address particular cases, whether created by Richardson's method of specification, balancing and weighting of principles, or a combination of these, and not necessarily to specification in the explicit sense described by Richardson. At least sometimes—and probably quite often—persons do make decisions based on principles without an explicit specification, and it is possible that at least some of these are derived from coherent systems. What follows should apply these systems as well as to explicitly specified systems.

---

[63] One might argue that this is an explicit specification, because it tells part of "in what circumstances" non-maleficence trumps beneficence. Specification seems to be more rigorous than this, but if one is willing to accept this as a specification, that is fully compatible with what follows below.

## A Further Difficulty

There remains an important problem that is as yet unresolved. Though it can resolve the first objection from Clouser, Gert, and Green above regarding the principles being merely vague chapter titles, specification still leaves a principle-based theory with something like the chapter title problem. Though a specified principle is no longer a chapter title, it may, after a lengthy period of specification, become something like a chapter itself. That is, the principle of "respect for autonomy" is, in the beginning, short, general and vague; it becomes much more explicit and specific after it is specified somewhat. But after much more specification (remembering that specification is a presumably unending and cumulative affair) the principle can become long and unwieldy, with many different codicils regarding various interactions with other principles, particular forms of application in types of cases, and the like. The "principle of respect for autonomy" could become, itself, a chapter-length analysis of various cases and conflicts, and could not really be used as the principles are meant to be used—e.g., something which could be reasonably easily grasped and understood and employed by persons as emergent cases arrive. If a true understanding and utilization of the principles requires grasping and holding in one's mind from one to four chapter-length dissertations on the topic of the principle, it may well be the case that persons will simply revert to using the "abbreviated" format and consider only "the principle of respect for autonomy" when a problematic case arises, and the principle would then serve as a chapter heading and nothing more.

This difficulty differs from the problem Clouser, Gert, and Green raise. Their objection was that the principles could do no more than serve as chapter titles to a discussion of various loosely connected and potentially contradictory thoughts about the topic. The problem they raise is not that the principle, as stated, is brief, but that "if the principle is not a clear, direct imperative at all, but simply a collection of suggestions and observations, occasionally conflicting, then [an agent] will not know [1] what is really guiding his action nor [2] what facts to regard as relevant nor [3] how to justify his action."[64] But even if persons begin to use the brief form of the principles as a means of reference to the lengthy and unwieldy fully specified form, they would not be referring by that to a "principle that does not tell the agent what or how to think, or how to deal with the value in a particular instance",[65] but rather to a lengthy discussion that gives precisely that guidance.

The concern that may still remain is *not* that the fully specified principles would become too unwieldy to use easily, or even to use at all. This seems rather weak as an objection: complexity ought not to be a legitimate criticism of attempts to explain the moral world, which is itself often complex. That persons can fail to use the principles properly by simply considering unspecified principles instead of the more fully specified versions is a concern in how the principles are employed, not a fundamental problem with the basic concept of principles.

---

[64] Clouser and Gert, "A Critique of Principlism", pp. 222–223.

[65] Clouser, "Common Morality", p. 223.

Rather, the more important concern that grows tangentially out of this worry, and which has been noted above in *Two Versions of the Common Morality* and *Specification and Reflective Equilibrium as a Response*, is that the specified principles may lose the nearly universal appeal that the general unspecified principles have. This is a major concern that can prevent the specified four-principles approach, as defended by Beauchamp and Childress, from being an acceptable means of resolving moral conflicts in a pluralistic context, because the ideal of reflective equilibrium to which they appeal cannot serve as a means of defending these resolutions as justified in that context.

## Specification and (Lack of) Universal Agreement

Clouser, Gert, and Green seem to critique a moral "system" of a few vaguely defined, general moral terms without the capacity for consistently or justifiably turning those terms into practical rules of action or judgments in specific cases; what the four-principles approach actually is, especially with principles being specified in the context of seeking maximal coherence, is little like that. Yet there was some value to the approach that Clouser, Gert, and Green critique; vague, unspecific principles of beneficence, respect for autonomy, nonmaleficence, and justice are more likely to be universally held to be morally important by most reasonable moral persons than are specific ones. The general principles would not derive their justification from that agreement, but by being agreed upon they would not serve, themselves, as a source of heated debate, even if the decisions based upon them might. Similarly, by being so vague as to not make any consistent specific judgments, the former approach also avoids making any controversial judgments.

But this is not what Beauchamp and Childress have done with their principle-based account. Rather, they have provided a means of creating specific, detailed principles and intricate levels of interaction, all of which is very much capable of deriving very specific conclusions and rules of action in particular cases. When this is done, the process of specification within the context of seeking reflective equilibrium commonly results in different persons, groups, or institutions rationally holding different, potentially mutually conflicting, sets of specified principles. In cases where participants in a decision have rational differences regarding specified principles, which may be common, they may not be able to appeal to a shared notion of the principles in the context of a justification of any action.

To see this, note the similar problem raised above in *Two Versions of Common Morality*, which is that the same words can and do have different meanings and entailments for different persons.[66] Thus, when one appeals to respect for auton-

---

[66] Engelhardt, *Foundations*, 2nd ed., pp. 56–58; Engelhardt, H. Tristram, Jr., and Kevin William Wildes. (1994). "The Four Principles of Health Care Ethics and Post-Modernity: Why a Libertarian Interpretation is Unavoidable", in *Principles of Health Care Ethics*, Raanan Gillon, ed. New York: John Wiley and Sons, pp. 135–147.

omy, one person may understand that appeal as a maximizing of the value of the liberty that each person has, and might thus sometimes require a violation of that liberty in order to preserve more of it later. Others may hold it to be related to the right to control over one's person that all should have, even at the cost of losing future control. Kantians might hold that one can only truly respect autonomy by defending action only when the actions in question are not chosen under influence of heteronomous inclinations; others might argue that respect for autonomy requires that persons support, or at least not oppose, any unforced choice of any action by a rational agent, whether heteronomous in Kantian terminology or not. "Respect for autonomy" is not clear because what respect entails and what autonomy is can differ from person to person. The same is true of the other principles, as well. Rawlsians will surely not understand the same thing by the term justice as will Nozickians, and their understandings of what is required in order to fulfill the requirements of justice will therefore also be different.[67]

The problem in general is that though the common morality, whether version 1 or 2, contains these crucial *terms*, persons may well not share the same meanings of those terms or norms. Thus, though there may be agreement on the acceptability of the four terms or phrases, beneficence, respect for autonomy, nonmaleficence and justice, there may not be agreement on what they mean and entail. Agreement on their meaning will not exist without prior agreement on a fair number of fairly important moral notions—for example, whether justice is served by seeking a just end-state or only by seeking just principles of acquisition and transfer. For reasons suggested below, this problem will often be increased rather than relieved by specification.

The problem is also exacerbated when one notes that features other than the principles, per se, can be germane to moral analyses. Any analysis of a case of commercial surrogate motherhood, for example, must involve views of the notion of human dignity and the purpose and value of reproduction and sexuality.[68] Any analysis of an ethical debate about utilizing fetal tissue, or animal experimentation, necessarily involves particular understandings of who and what has moral standing. To whom or what ought one be beneficent? These are, of course, questions of scope, but they are still important for those who wish to determine how the principles can be specified and understood. Though the four principles can be used to approach these issues, different understandings of these views are common, and the principles, to the extent they can be used to address these important ideas, must necessarily apply differently for persons who hold those different views. So, from this and the above, it would seem that the principles, even if agreed upon in some basic form and even if applicable to bioethical problems, cannot be universally understood to have a

---

[67] At their hearts, different concepts cannot differ too dramatically—e.g., one cannot legitimately understand "respect for autonomy" as "providing the most good for the most persons" or "justice" as "being as unfair as possible"—but even early specifications of legitimately different viewpoints will differ, sometimes very significantly. See *Differing Specifications in a Pluralistic Society* below for further discussion of this.

[68] Engelhardt and Wildes, "The Four Principles and Post-Modernity," p. 145.

common meaning and field of application. Different persons hold the principles to mean significantly, though not totally, different things, and no appeal to shared principles can be used to resolve moral conflicts involving those different persons to the satisfaction of all.

Yet this critique cannot be entirely correct. There seems to be something very right, to which Beauchamp and Childress appeal, in the notion of a common morality that includes the four principles. Consider the example that Michael Walzer uses to begin *Thick and Thin*:

> I want to begin my argument by recalling a picture.... It is a picture of people marching in the streets of Prague; they carry signs, some of which say, simply, "Truth", and others, "Justice". When I saw the picture, I knew immediately what the signs meant - and so did everyone else who saw the same picture. Not only that: I also recognized and acknowledged the values that the marchers were defending - and so did (almost) everyone else.[69]

Walzer seems to be appealing to about the same basic intuition to which Beauchamp and Childress appeal: there is a shared notion of truth and justice that all moral persons have, though there are vast differences at the theoretical level. Nozickians and Rawlsians would surely be able to understand the meanings of the signs of these marchers in essentially the same way, and indeed could both join in the march. This, then, may be what the principle of justice picks out: this common understanding of the term that transcends the deep conflicts between differing interpretations. The notion that Walzer assumes, that moral persons all have this common understanding, is precisely the intuition that Beauchamp and Childress have hit upon.

As Walzer notes, the marchers were not marching in favor of a coherence theory of truth, nor a correspondence theory: they were marching in order to make the government and the media stop blatantly lying to them. In that sense, they shared a concept, though not a theory, of truth. Similarly, they were not marching for a particular theory of justice, but justice in a more general sense: impartiality and a lack of arbitrariness in the exercise of the laws and governmental power—what Walzer calls "common, garden variety justice."[70] Surely every plausible theory of justice shares this notion; perhaps this shared concept is what the principle of justice is meant to entail. Similar arguments could be made for each of the other principles having a commonly held, garden-variety meaning.

This, then, might be what Beauchamp and Childress can argue: there is a commonly held, garden-variety level understanding of each of the principles in the common morality that can serve to begin moral reflection. This understanding of each of the principles is neither confused nor unshared at the outset, and thus can serve as a means for all to have shared moral content to begin specification and development of these moral principles.[71]

---

[69] Walzer, Michael. (1994). *Thick and Thin: Moral Argument at Home and Abroad*. Notre Dame, IN: University of Notre Dame Press, p. 1.

[70] Ibid., p. 2.

[71] Personal communication with Tom L. Beauchamp, July 2000.

But it does not follow, and Beauchamp and Childress do not argue, that specified principles must therefore be shared by all serious moral persons. While any plausible moral theory ought to share this "garden-variety" notion of justice—meaning, in this case, impartial law enforcement and an end to arbitrary arrests—and thus there is an agreement on some part of the meaning of justice between persons, the four-principles approach allows for multiple specific and substantive material principles of justice. If it did not, it would run afoul of the accusations of overgenerality made by Clouser, Gert, and Green above. Walzer's garden-variety justice as a principle would hardly be useful in resolving questions of, for example, whether a surrogate mother may justly invalidate a contract and keep the infant she delivered, or, indeed, whether surrogate motherhood contracts are ever just.

Beauchamp and Childress do not restrict the principles to this limited form, of course: they hold that the common morality, garden-variety understanding is merely the starting point from which one analyzes and specifies the principles. At this point there is a problem for anyone hoping to use the four-principles approach to resolve moral conflict in a pluralistic society. One can develop the principles and specify them in particular situations to make them more precise and to expand them beyond their garden-variety beginnings; this is necessary in order to make the principles useful for resolving moral problems. But when one does this, one leaves behind the universally shared principles and develops them into principles that are not universally shared in a broadly diverse pluralistic society. Beauchamp and Childress agree that a specified principle of justice will very soon expand beyond that "garden variety" justice, and in so doing, it may open itself up to the possibility of rational disagreement. This keeps a four-principles approach from being sufficiently universal to be accepted by all or nearly all. The problem with this for resolving problems in a pluralistic society is thus that one cannot count on having a system that is both broadly shared and capable of resolving moral conflicts. Given that different persons specify the principles in the context of rationally differing moral worldviews and/or theories, one can be almost certain that different specifications of the principles will occur. One therefore cannot expect to make the four-principles system into specified principles that can always both (1) be justified to all or nearly all persons in a pluralistic society and (2) resolve moral conflicts.

A theory with well-specified principles will be able to resolve problematic cases and justify those resolutions by means of more explicit specification, which is itself justified by the greater coherence of the more specified system over the less so. But it is precisely this specification and moving away from the general to the specific that can and, given the broadly differing conceptions of morality that exist in a pluralistic society, most likely will, move the resultant system out of the realm of (nearly) universally accepted to a far more narrowly accepted system which may not be able to justify its decisions to persons with differing moral views.

## Differing Specifications in a Pluralistic Society

When the suggestion is made here that shared specifications are not likely in a pluralistic society, I mean a fairly strong version of the claim: it is sufficiently unlikely that significant specifications of moral principles will be shared across a pluralistic society that it cannot be assumed in any moderately difficult case that those considering the case share specifications. It is often true that specifications are shared by some persons, sometimes for differing reasons, the ramifications of which are explored later. But, if the arguments here are correct, one cannot *assume* this agreement in any given case, and must show it to be explicitly true in the specific case at hand. This will be true for full sets of varied specified principles as well as for any particular specification within a given set of principles.

The reason for this is that specification does not and cannot determine a unique means of specifying a principle in combination with the fact that the wide variety of rational but different moral views in a pluralistic society make it extremely likely that different specifications will occur. The problem is this: in developing a specified system, many particular specifications will occur that may not, at the time, result in a significantly diminished coherence either at the time of taking the step or in the perceivable future. These steps are neither motivated nor determined by coherence, then, but rather by other values and evaluations of the person making those steps. Because different rational persons value different things, and value the same things at different levels, there can and will be different specified systems of principles that are similarly coherent, but justify significantly different actions in particular cases. Specified principlism can create a system that can give explicit instructions in particular cases, but it does not do that without making a number of choices along the way that will tend to separate moral strangers and some moral friends, insofar as they hold somewhat different views, keeping them from necessarily agreeing upon any particular specification, rule, or judgment. It will also distinguish the principles from each other such that, eventually, the Nozickians and the Rawlsians mentioned above will indeed have principles of justice that differ dramatically, though they both still agree on what the marchers in Prague desired.

The argument for this is straightforward: if one begins, as Beauchamp and Childress do, with unspecified principles drawn from the common morality, and has a significantly specified and ordered set as one's goal, the journey must begin with a single step. In this case, that is a first specification. But how might that specification be justified? It cannot be justified in terms of coherence with the remaining whole—that whole is not yet coherent enough itself to be noticeably better or worse with any particular first step. With what will its coherence be judged, then? If at all, it is judged as being more or less coherent with the individual's other considered moral beliefs, both theoretical and pre-theoretical. It is indicative of the importance of these considered beliefs in driving the specifications that among them will be the pre-theoretical analysis of who and/or what has moral relevance, and why. Basic beliefs such as "Humans alone have souls, and thus have a privileged moral status," or "Any sentient being is comparably morally relevant to any other sentient being," are two obviously contradictory examples of beliefs along these lines. Until the system of specified beliefs becomes large and detailed enough to have a coherence of its own as a system, there will be nothing in the system itself to guide its growth. By the time the system is complex and complete enough to be a significant feature in a wide reflective equilibrium, it will very likely be developed in ways that will not be shared by all, because the disparate beginnings which initially guided the varied systems were not shared by all.

Persons' differing basic values and differing evaluations of those values, as well as differing views as to the appropriate targets of those views, will drive much of the critical initial specification. The common morality cannot fully guide these decisions, as the specifications are explicitly expanding on the understanding in the common morality. The common morality sets limitations on what specifications are initially possible, but cannot provide sufficient guidance to determine a conclusion as to which specification is actually best. So, though a specified system can hold some judgments as wrong and others as correct in particular cases, turning the shared common morality intuitions into a specified system can and will surely occur on numerous parallel, but different, paths. In one sense, this is hardly surprising; given the wide differences in moral values that people in general hold, there is some reason to think that they really may not share so much in common at the middle level.

Moreover, this differentiation can continue to progress after the system of specified principles becomes complex and coherent enough to serve as grounds for measuring the coherence of a particular specification. First, the theoretical and pre-theoretical intuitions that would have guided the original specifications will be reflected in the system of specified principles one now uses to test the coherence of a new specification; as well, those theoretical and pre-theoretical intuitions will normally also still be present themselves. If a particular specification contradicts those concepts, that itself will provide a reason to reject it, though not necessarily an overriding reason. These concepts will still be a part of that system in which one would normally seek wide reflective equilibrium. Unless there is a reason given for why one must ignore those other moral considerations not explicitly encompassed by the four principles, there seems to be no reason for persons to ignore them in their moral reflection.

Even if this latter concern is disputed, the initial development of the specification of principles can be so different between persons with different rational moral beliefs that there is little likelihood that they will arrive at the same set of specified principles. Thus, specification of principles, even with the constraints of wide reflective equilibrium, is not likely to produce a single, most coherent set of principles; rather, it is likely that a pluralistic society will include many similarly coherent different sets of specified principles.

One could, at this point, agree with all of the forgoing but argue as follows: "Yes, there may be different systems built up from the same basic roots. And at some point early on, they may all seem to be approximately similarly coherent. But not all of these systems will remain so coherent as others, when enough judgments and rules and specifications of the principles are made, and thus, eventually, we will be able to determine those that cannot keep up, and remove them from contention, until we have only the most coherent system. The beginnings of the specification are by necessity somewhat subject to individual whims and ad hoc, but the constraints of wide reflective equilibrium are not, and thus, eventually, only the most coherent system will remain."

This line of argument seems too idealistic. Though it is true to some extent that some specified systems will be obviously less coherent than some others, in order to achieve a system by which moral strangers can resolve moral conflicts, one will need to winnow down the systems until only one system, or at most a few very similar systems, remains. It seems unlikely that the human capacity for measuring coherence can achieve this. Consider the differing views represented in the Terri Schiavo case discussed earlier.[72] The members of these two groups could not arrive at the same system of specified principles, because what they value and how they value it will influence their development of that specified system. Yet neither group is obviously more incoherent than the other; each is simply more coherent with different sets of views. The same situation would be true of Singerian vegetarians confronting Eastern Orthodox Catholics regarding the consumption of meat. Neither has an incoherent set of beliefs, but their belief sets are coherent with different claims about the relevance of sentience and of species. The Singerians could argue that all sentient beings are worthy of similar moral respect, while the Eastern Orthodox Catholics could argue that eating meat is a gift from God that it is wrong to refuse on moral grounds. Each group would surely locate the subjects of beneficence and nonmaleficence, and probably respect for autonomy and justice as well, differently. But it is unlikely that either view can be shown to be more or less coherent than the other—for one thing, too many of the crucial relevant claims are unverifiable.

The above argument that wide reflective equilibrium and coherence can eventually show one set of specified principles to be superior to all others (which is a possible argument, though not one which has, to this author's knowledge, been made) is rendered unlikely, though not disproven, by this line of argument. But since a rigorous proof of the claim is not possible—as it would necessarily involve comparison of all possible different sets of beliefs—leaving the discussion at the level of raising significant doubt is appropriate. It has also been shown at the other end of the picture that there do exist significantly different, large sets of beliefs that cannot be shown to be differently internally coherent, each of which could well be differently specified sets of principles.

## Justification of Actions on the Principles Approach

Assuming that everything argued above is correct, one might ask what a principle-based account might be able to do in a modern, pluralistic society. It cannot be a means for *all* rational persons to justifiably resolve *all* moral conflicts, but perhaps it can still be useful in determining actions to take in morally problematic cases. And a first answer might be given that it could still do a lot, since there is a wide general agreement on the principles and at least some of their specifications in society, though there are dramatic moral differences among some persons, and though

---

[72] See Chapter 1, *Moral Disagreement in a Secular, Pluralistic Society.*

there are some few who do not share even the basic specifications of the principles. Not everyone will agree with the specified set of principles that Beauchamp and Childress derive, but one could argue that most people will agree with enough of them to resolve most or nearly all moral difficulties in medicine. Thus, one could argue, despite what is said above, the four-principles approach could usefully be employed to find solutions to moral problems that could be justified to most persons in a pluralistic society, and thus could be used in many cases to resolve moral conflicts satisfactorily. The above could be seen as a relevant, but relatively minor, point to consider, but hardly a major criticism of an attempt to use a principle-based approach as a means of moral conflict resolution in a pluralistic society.

This initially seems a plausible response. Principles may be able to address cases 2.1–2.3 from the last chapter, and they seem a plausible way to approach other significantly difficult moral cases, as can be shown by analysis of a case from the Georgetown University Hospital some years ago:

**Case 3.2: Refusal of Blood Products by a Pregnant Minor** A 15-year-old Jehovah's Witness who was 4 months pregnant and had a two-level spondyloptotic cervical spine fracture was transferred to the Georgetown University hospital 3 weeks after being injured in an automobile accident in the hopes that her quadriplegia could be alleviated or removed by an operation on her injured spine. In addition, she was severely anemic and refused transfusion of blood products. The injury was apparently unprecedented in the literature, and the additional complexities of the case (her pregnancy, her minor status, and her anemia and refusal of blood products) made it more difficult to address. With a careful ethical and technical analysis, including use of the four principles, a solution was arrived at that satisfied all the participants, avoided blood transfusions, and treated the patient's spinal injury without further risk to the fetus. The autonomy of the patient was respected, and a beneficent result was provided to her without causing or significantly risking harm to the fetus.[73]

The major ethical challenges in this case were seen to be determining whether a pregnant 15-year-old was competent to make a life-threatening decision and assessing the ethical status of the fetus. It was argued that the "moral competency" of this patient to assess her own situation and make a valid decision in this case was adequate for her to have a sense of autonomy that ought to be deserving of respect: after numerous interviews she was judged to be intelligent and mature for her age and capable of understanding the risks and benefits of her decision. Furthermore, various factors opposing the labeling of her choice as autonomous were judged to be impacting her decision, but not directing it. The team addressed and rejected the possibility that her decision was guided mainly by guilt based on the social and

---

[73] Feigenbaum F, Sulmasy DP, Pellegrino ED, and Henderson FC. (1997). "Spondyloptotic Fracture of the Cervical Spine in a Pregnant, Anemic Jehovah's Witness: Technical and Ethical Considerations." *Journal of Neurosurgery* 87(3):458–463. This case presentation is my paraphrase of the facts given in this article. Though I argue that specification of the principles is a large part of what is allowing the resolution of this case, one might question whether there is any actual specification going on, or simply balancing of the principles. In this I agree with Richardson that, if not any case, then at least many cases of balancing can be made into a case of specification; here it seems that all that would be needed to do so would be a reasonably clear analysis of the reasoning used to make the decisions regarding the appropriate understanding of the various principles here. See Richardson, "Specifying, Balancing, and Interpreting Bioethical Principles," p. 300.

religious sanctions she had already received by becoming pregnant outside of marriage, and also the possibility that she was deciding based on a desire to please her Jehovah's Witness parents. Thus, she was held to be autonomously guiding her therapy.[74]

It was also argued that respecting her choice should take precedence over providing a medically beneficent prognosis for her, despite the fact that she was classified under District of Columbia law as a minor, and District law held that withholding life-sustaining treatments from a minor for religious reasons is child abuse. On the grounds that she was morally competent to make such a choice and had adult level decision-making capabilities, her choice was determined to be valid and important.[75] In the language of principles, one could describe this as arguing that respect for the autonomous choice of one capable of making a good autonomous choice, regardless of age, generally ought to take priority over providing a beneficent act for that person by overriding that autonomous choice.

The welfare of the fetus was also at risk, and since the patient wished to bring the fetus to term, if possible, several persons questioned whether beneficence to the fetus ought to outweigh her decision to risk the life of the fetus by refusing blood products. She might be able to choose for herself what treatment she wanted, but she could not impose religious views on the fetus that threatened its health. Since the main risk to the fetus was determined to be from a catastrophic bleed during the operation, which could probably not be reversed with blood products in time to save the fetus, her autonomous choice to refuse blood was thought to still be more important.[76] It was also noted that the fetus was not yet viable, and thus adjudged to be legally subject to the patient's authority. Thus, it was argued, her autonomous choice to refuse blood products ought to be respected, though that decision provided some risk of harm to both her and her fetus.

Now, one might take this as a good sign of the power of the four principles, carefully specified, to address moral problems and arrive at solutions justifiable to all relevant parties. After all, there was significant moral disagreement involved in the case, yet it was resolved in a fashion which all persons took to be justified, with respect for her autonomy, beneficence, and nonmaleficence guiding the actions in the case. The Catholic hospital's concerns for fetal life were accommodated, as were the Jehovah's Witness's concerns for avoiding blood transfusions; the surgeon's concerns for ensuring a good prognosis for the patient were addressed, as were the concerns of all as to the status of a 15-year-old's capacity to choose risky medical procedures, whether her pregnant status made her concerns more viable, and whether her pregnant status meant that the health and life of the fetus might impact whether her choices for her treatment were acceptable. Yet, the attempt to

---

[74] Feigenbaum, et al., p. 461.

[75] Ibid., pp. 461–462.

[76] Ibid., p. 462. What the health care team was considering was initiating a transfusion to protect the fetus in the case of a catastrophic bleed; it isn't clear from the article whether this risk could have been better alleviated by using blood products throughout the procedure (meaning that the risk of a bleed might have only been severe if the transfusion had to be begun at that point).

justify that decision shows the difficulty in understanding this case as an indication that the principles can provide a means for conflict resolution, and the depth of the problem of varying specifications of principles argued for above.

The justification of the decision, as Beauchamp and Childress would argue, should come from the innovative character of the specifications and the coherence of the system of specified principles, in addition to the shared considered judgments regarding the principles themselves.[77] The justification for actions derived by appeal to the principles can be found on grounds of a reflective equilibrium reached by constructing principles and rules from agreements in the common morality, and then specifying them.[78] The justification of the resultant theory depends upon the coherence of the system, but also, importantly, upon the considered agreement on the principles towards which they find that competing moral thoughts converge when seeking practical solutions.[79] Mere coherence of a system itself cannot suffice to justify a system as a moral system, else one would be stuck in the unacceptable position of affirming the Pirate's Creed or a coherent Nazi system—if such a thing is possible—as valid ethical systems.[80] The decisions that can be made by employing principles depend in large part upon the convergence of agreement in the common morality shared by reasonable moral persons for their justification.

The objection that is provided below can be pithily, but not wholly inaccurately, stated as: insofar as there is shared agreement on the principles, they cannot be successfully appealed to as justification for resolutions to moral problems; insofar as principles can serve as considered judgments as those judgments are meant to function as part of a reflective equilibrium, they are often unshared. The problem is, as argued above, there is no convergence of shared considered judgments to which Beauchamp and Childress can appeal for justification. The principles may be shared at the level of vague and general moral terms with minimal content, but at this level of generality the principles cannot serve as a useful shared set of considered judgments. A considered judgment is one "given under conditions favorable for deliberations and judgment in general"[81] and a judgment "in which our moral capacities are most likely to be displayed without distortion."[82] The nature of the principles as they are potentially shared in the common morality is not one in which they are likely to be displayed without distortion, precisely because their generality leads to their needing further definition and specification before it is understood what they mean. That is, a claim that "Beneficence is important" is too vague to be a judgment displayed without distortion until one has a much clearer definition of what the central term, beneficence, means. Insofar as the principles are compatible with both the claim "One ought to do what one recognizes to be good for another," and "One

---

[77] Beauchamp and Childress, *Principles*, 5th ed., p. 400.

[78] Ibid., pp. 398–400.

[79] Ibid., pp. 376–377.

[80] Ibid., p. 400.

[81] Rawls, John. (1971). *A Theory of Justice*. Cambridge, MA: Harvard University Press, p. 48.

[82] Rawls, *A Theory of Justice*, p. 47.

ought to do for others what they recognize to be good for them," it is not specific enough to be properly described as displaying anything about beneficence without distortion. The vagueness of the "garden-variety" principles shared in the common morality presents the principles and human moral capacities at their fuzziest.

Though the garden-variety notion of justice to which Walzer appeals, and to which it is argued above that Beauchamp and Childress can appeal as being shared in the common morality, represents a shared notion that one ought to consider fairness and equality in one's actions towards others, this is not a clear perspective of what justice is; rather, it is the opposite. One understands by it some very basic things, such as "Actions which are known to be unfair generally ought not be performed," but it tells little or nothing about why one ought not be unfair, nor what sorts of actions are unfair. These sorts of questions cannot be answered until the principles are specified; but specified principles, because they are more likely to be unshared the more they are specified, cannot in many cases be commonly shared considered judgments to which one can appeal for justification of a decision. What is more, the language of considered judgments suggests that these judgments ought to be carefully thought out before they are used in an attempt to find reflective equilibrium, which raises another problem: once considered carefully, these judgments may no longer be shared, for careful consideration will lead persons to ask just those sorts of questions required to clarify the general, vague terms. Once one asks and answers questions of why a principle is moral, and what sorts of things run counter to that principle, one is getting into the territory where specifications may occur that are not necessarily—and, as argued above, may often not be—shared. Because only principles which are specified at least somewhat beyond the garden-variety level are clear statements best suited to expression of our moral capacities without vagueness and distortion, only such specified principles could be considered judgments; but if the arguments above are correct, they are not universally, or generally, shared.

Even if the principles in some general form shared in the common morality are valid considered judgments, there is a further problem with justifying the conclusions of a principle-based account even partially by appeal to the considered judgments of the common morality. The principles that are used to actually address problematic cases are specified ones, not the garden-variety ones in the common morality. A decision made in accordance with a carefully specified, fairly complex principle would be justified, not by appeal to that specified principle, but by appeal to the general principle in the common morality. It seems strikingly odd to appeal to a vague and unclear "garden-variety" notion of justice for a justification of a decision based in, say, a fairly clear and explicit notion of utilitarian justice—yet if defenders of principles are to appeal to shared common-morality claims at all for justification of a claim, this is what a utilitarian making that decision must do. On such grounds, the same vague concepts could be grounds for justification of opposing actions derived from dramatically different concepts of the same principle. One could justify transfusing the fifteen-year old patient in Case 3.2 on grounds that it would be beneficial to her if beneficence is understood to mean, "Do unto others what you recognize to be good for them." Or, one could justify refusing to transfuse her on the grounds that it would be beneficial to her, if beneficence is understood to mean, "Do unto others what they recognize to be good for them." In such a case, it seems far more important that persons disagree on the specific norms than that they agree on some vague but non-action-guiding version of them—i.e., the shared, but general, form of the principles. Shared garden-variety principles would be, at best, viewed as "starting points for deliberation" needed to get to the specified but unshared versions, and would not be useful or used in a justification once the more specified version was derived.

This can be seen in Case 3.2 above, which was resolved on grounds of claims which, to the extent that they were derived from principles, were derived from rather different sets of specified principles. There may have been adequate agreement about whether the patient was adequately autonomous in her decisions to have that autonomy respected so that all persons involved may have agreed that it was appropriate to specify respect for autonomy to allow her to be treated as an autonomous agent; similar agreement was lacking regarding the other principles. The importance of this can perhaps be seen by focusing on the central issue of concern in this case. The patient held that it would be significantly beneficent to her for her not to receive a blood transfusion, since she believes that a blood transfusion would interfere with her prospects for eternal happiness, while the physicians believed that the transfusion would be beneficent for both her and her fetus. The physicians eventually held that the respect for the patient's autonomy that they must show the patient, combined with the fact that her fetus was not of sufficient age to require that they take it into more direct consideration, required them to refrain from performing this beneficent act on her against her will. They each agree, for very different reasons, that the patient ought not be transfused. But to appeal, even in part, for a justification of this solution to the belief shared by each that beneficence and respect for autonomy are both important moral principles seems strange at best. Their notions of what beneficence and respect for autonomy, and what they entail, are significantly different in this case, and a justification of their agreement on an action on the grounds of their shared common morality agreement on the principles—when, in fact, it was an overlap in significantly different principles that actually determined that action to be best—seems to miss not only the reason for the choice of that action, but also the justification they might actually have for their decisions. Two different systems of moral claims, each of which may well have little to do with each other, each recommend the same action in this case; but this action cannot plausibly be justified by any appeal to reflective equilibrium that these two parties have reached. This action also cannot be justified by an appeal to a shared common morality because it was derived from particular moralities developed from unshared branchings from that vague, unhelpful common morality.

The justification of actions derived from application of the principles *via* an appeal to the common morality plus the coherence of the system is reduced to mere coherence, which Beauchamp and Childress have rejected as adequate grounds for thinking a system is a moral system. Thus, they cannot hold the conclusions of principle-based deliberations to be justified on the grounds they admit for justification; though a solution has been attained, it is as yet unjustified.

What this shows, both for and against the usefulness of a four-principles approach, is interesting. The four-principles approach cannot successfully appeal to the common morality, or indeed any commonly shared claims, to be a means for persons in a pluralistic society to be able to resolve, with secular moral justification, medical ethical problems. Yet it seems that principles can be used to resolve some, and perhaps many, such problems as long as there is adequate moral agreement on an adequate number of specifications of principles, or at least agreement on basic concepts adequate to justify an agreement on specifications of principles. There was, after all, a solution to the difficult Case 3.2 above that morally satisfied all parties to the case, with the assistance of an appeal to principles. The question is now how to make a justified decision.

A hint to a possible solution comes from noting that the resolution actually derived in the case does not satisfy all morally serious persons. At a presentation of this case in a medical school ethics class, objections were raised by various physicians in the audience. They argued that the surgical resolution of the spine fracture was not the best care possible for the patient, though it was generally agreed that it was the best care possible that could be provided without blood products. The treatment provided involved a good chance of reducing or eliminating the patient's quadriplegia, but not, to the best of their knowledge, the best possible chance of that. The surgical procedure involving blood products was expected to provide a better chance at relieving her symptoms. These physicians felt that appropriate beneficence required providing the patient with the best possible physical prognosis, even if that involved providing blood products, because her minor status diminished the relevance of respecting her choice. In other words, the solution that was derived in the case did not satisfy all reasonable persons, and part of the reason for this was that they specified beneficence and respect for autonomy differently, and may have, through those differing specifications, ranked their importance in the given case differently.

At the same time, the decision attained in Case 3.2 was sufficient for the physicians and patient in question to satisfactorily perform an operation and believe that the decision to do so was morally justified. It was not, perhaps, justified or justifiable to all persons, but it was justified in some important sense: it was justified to all the active participants in the case.[83]

One also does not know whether the persons involved shared the specifications of principles, or whether there was an overlap in differing sets of specified principles. It is likely that there was only an overlap, since the patient held differing beliefs about the beneficence of avoiding transfusion than did the hospital; yet this overlap seems sufficient to allow the group to make and justify their decision. They were able to justify their decision; but this justification cannot be shown on the grounds which Beauchamp and Childress give for justification of resolutions by the principles.

---

[83] For a further discussion of this kind of justification, see Chapter 5, *Moral Acquaintanceships and the Mini-Culture of Medical Cases.*

Perhaps this suggests that, though the attempt to derive a system of specified principles justifiable to all by appeal to the common morality cannot succeed, there can be a way to use specified principles to resolve specific cases, wherein the principles in question need not be justified to all persons, but need only be justified to persons who are involved with a particular case. Perhaps decisions can be justified, not by sharing the same coherent system or partial system of principles, but rather by an overlap of coherent systems. This is what has happened in Case 3.2, where all of those involved agree that a morally justified resolution had been achieved, though some others who reflected upon it at a later time challenged it. The specification of autonomy rights vs. beneficence towards the minor and her fetus were addressed in a fashion that was agreed upon, and held to be justified, by patient, family, and the health care team, though not necessarily all for the same reasons. Though this does not mean the same specification would be acceptable to all, it may well mean something important—the conflict was resolved and all persons directly involved held that the resolution was morally justified.

This would show the ability of principles to resolve moral problems in the context of sufficient agreement on sufficient principles or prior moral claims. This resolution was not justified by a joint appeal to common morality, but rather by appeal to different values that were held or weighted in different fashions by the various persons involved in the case. The specifications which each person could justify overlapped enough to allow them to agree on a solution. In this way, the method of specification of principles resolves the problem, to the satisfaction of those directly involved with the problem, but the justification of that resolution appealed to particular, not common, understandings of morality. This leaves open the question how the resolution could be justified by appeal to these different but overlapping grounds; some possibilities along these lines are discussed in the final chapter.

However, there is another commonly referenced method of resolving moral conflicts in a pluralistic society: casuistry. If, as some of its proponents argue, it can appeal to a common agreement that is broader than the common morality as understood by Beauchamp and Childress, casuistry may be the best approach to take in order to address and resolve these problems. In the following chapter, it will be shown that casuistry turns out to have many of the same strengths and weaknesses of principle-based theories, and will thus not be able to accomplish all that could be desired of it. The final chapter will ask on what grounds modern societies could resolve moral problems in a pluralistic context.

# Chapter 4
# Casuistry in a Pluralistic Society

The medieval practice of casuistry, the practice of case-based reasoning by analogy to similar cases, has been revived as another means of addressing moral conflicts in medical cases.[1] The renaissance of medical ethical casuistry came about when some bioethicists began to note that, very often, decisions made about particular problematic medical cases were more commonly agreed upon than decisions about moral theory. Whereas discussion of moral theory often led to conflict and disagreement, discussion of particular cases or particular specific guidelines often led to an agreement on a morally correct course of action.[2] Casuists purport to resolve morally problematic cases by reference to the salient issues of the case and comparison to other previously decided cases, without reference to an overarching moral theory. This last is very important: if this is successful, casuistry can enable one to resolve moral conflicts in a pluralistic society without any need to refer to the rationally contested moral theories that presumably separate the different parties to a particular problem. A casuistry that could actually accomplish its goals of resolving problematic cases in such a way that all persons can recognize the solution as morally right would thus be an ideal means of resolving moral conflicts in a pluralistic society.

After a brief description of the functioning of casuistry, I argue that the method depends on an underlying theory or set of moral claims that cannot be presumed in a pluralistic society. Casuistry needs particular moral content to resolve the many problems encountered in medical ethics, and must presume such content even to derive some of the paradigm cases that are the building blocks of the theory. The means to which some casuists appeal to obtain this "higher-level" content runs into many of the same problems as were raised against principle-based theory in the previous chapter. After two possible means of seeking enough shared content to

---

[1] Jonsen, Albert R. and Stephen Toulmin. (1988). *The Abuse of Casuistry*. Berkeley: University of California Press; Strong, Carson. (1988). "Justification in Ethics", in *Moral Theory and Moral Judgments in Medical Ethics*, Baruch Brody, ed. Dordrecht: Kluwer Press, pp. 193–211; Jonsen, Albert R. "Of Balloons and Bicycles: The Relationship between Ethical Theory and Practical Judgment." Hastings Center Report, September–October 1991, pp. 14–16; Arras, John. (1991). "Getting Down to Cases: The Revival of Casuistry in Bioethics." *Journal of Medicine and Philosophy* 16: 29–51.

[2] Jonsen and Toulmin. *The Abuse of Casuistry*, pp. 16–20.

S.S. Hanson, *Moral Acquaintances and Moral Decisions*, Philosophy and Medicine 103, DOI 10.1007/978-90-481-2508-1_4, © Springer Science+Business Media B.V. 2009

make casuistry work are refuted, casuistry will be shown to be unable to resolve problems in a pluralistic society, since it must assume too much unshared content.

## How Casuistry Works

Casuistry is a practice of analyzing particular cases by reference to their similarities to and differences from other, comparable cases. Casuistry works by taking current cases and deciding how to resolve them by comparing them with other, previously decided, analogous cases.[3] It is, as one proponent describes it, "a case-based approach in which an argument is developed by comparing the case at hand with *paradigm* cases in which it is reasonably clear what course of action should be taken."[4] In principle, how this is done is very simple: one examines a given case and determines what sort of more easily resolved paradigm case or cases it most resembles; then one determines whether it has any relevant different features that entail that this case ought to be decided differently than the paradigm case(s). Jonsen and Toulmin describe the process of practical resolution as analysis of the particulars of the instant case and the derivation of a general warrant based on similar precedents. This provides a provisional conclusion about the case at hand, which is subject to various rebuttals based on exceptional circumstances, or is not subject to them if no such circumstances exist.[5] A good casuistic reasoner must understand the circumstances of the case at hand, must understand what paradigm cases are appropriate guides, and must understand the features that make the two cases different and how they may make the right action in the case at hand different from the paradigm case. This is elaborated in a case which shall be returned to later:

> **Case 4.1: A Jehovah's Witness's Children** The problematic case is a pregnant, near-term Jehovah's Witness woman who developed abruptio placenta (premature separation of the placenta from the uterus), requiring a C-section to protect both mother and infant. She agreed to the C-section, but with no blood transfusion. Following surgery, which delivered a healthy child, she developed further complications, requiring blood transfusion to save her life. She continued to refuse blood, with the support of her husband, who said that if his wife dies, he and other family could care for their seven children [ranging in age from four to seventeen], despite their financial insecurity [the husband works as a yardman and also receives welfare assistance, which would continue after the patient's death].[6]

---

[3] Though there are different approaches to casuistry that differ in some of their terms and finer points, all forms of casuistry work more or less in the same fashion. The differences between the various forms do not make much difference in actual practice. Note, for example, the ease with which Tomlinson puts Strong's casuistry into Jonsen's terminology (Tom Tomlinson (1994). "Casuistry in Medical Ethics: Rehabilitated, or Repeat Offender?" *Theoretical Medicine* 15: 5–20, p. 9) without any apparent change in the casuistic analysis or in the resultant decision. In any case, the basic descriptions of each method are vulnerable to the critiques herein.

[4] Strong, Carson. (1999). "Critiques of Casuistry and Why They Are Mistaken." *Theoretical Medicine and Bioethics* 20: 395–411, p. 396. Italics in original.

[5] Jonsen and Toulmin, *The Abuse of Casuistry*, p. 35. Some examples are elaborated on pp. 323–324.

[6] Tomlinson, "Casuistry in Medical Ethics: Rehabilitated, or Repeat Offender?", p. 9. He takes this case from Strong, "Justification in Ethics", p. 195. The accounts are identical, but for the fact that Tomlinson has used Jonsen and Toulmin's terminology and methodology.

The important concerns to be touched on seem to be "respect the wishes of a competent patient" and "protect the interests of dependent children." A good paradigm case will thus address similar conflicts of autonomy and benefit or harm to dependents, and should otherwise be similar enough to be appropriately analogous.

Strong compares this case with two different paradigm cases (which shall be referred to herein as Case 4.2 and Case 4.3), which are very similar to Case 4.1 "with regard to the patients' autonomy, the degree of physical harm to the patients [and the patients' dependents] to be prevented by the treatment, or the degree of psychological harm to the patients in overriding their wishes. ..."[7] These two cases also involve parents of young children refusing life-saving medical treatments, but differ from each other in minor but important ways. In Strong's Case 4.2, the medical details are similar to Case 4.1; in this case, if the parent does not survive, her children will very likely have minimal to no financial support and difficulties with familial and emotional support. In this case, a judge ordered forced transfusion. In Strong's Case 4.3, on the other hand, there is a large, financially secure, close-knit family structure ready and willing to provide all forms of support for the surviving children; here, a judge refused to grant the court order for transfusion. Because of the similarity of Case 4.1 to Case 4.2 in terms of the likelihood of harm to the surviving children, Strong argued that in Case 4.1, too, a court order should be sought to transfuse the patient if she cannot be convinced to choose transfusion on her own, because there is significant doubt as to whether the dependent others will be taken care of in this case. Since there are no other important exceptions, he argued, this case resembles the first paradigm and a court order to force transfusion should be sought.[8] Concerns for the children in that case were thought to override concerns for the autonomous choice of the parent, and so, by analogy, are the same beneficent concerns held to trump concerns of respect for autonomy.[9]

## Problems with Analogical Reasoning: Maxims and the Common Morality

This strategy is limited in its effectiveness in a pluralistic society. Casuistry functions by seeking more easily resolved cases that are relevantly analogous to the difficult case at hand. An analogous case is one which is relevantly similar to another case, though different in some less important ways; which features are relevant and which differences are less important are not defined by the case itself. Reasoning by analogy therefore must have four central features: a need for consistency between

---

[7] Strong, "Justification in Ethics", p. 205.

[8] Ibid.

[9] Strong does not now argue that forced transfusion is appropriate, but rather that in society now, perhaps contrary to when he originally wrote, it is inappropriate to force transfusion even in Case 4.1 or Case 4.2. The importance of this point is addressed below in the next section.

decisions, a focus on particulars and particular cases, operation without a central-
ized theory, but also principles that operate at an intermediate level of generality.[10]
Justification of the choice of a given paradigm as the appropriate paradigm because
it best resembles the case at hand, knowing which of given moral features of a case
are the most relevant, and so on, does not allow for a definitive result without an
appeal to more general justifying principles of relevance.

Normally, analogical reasoning working within the context of some basic
assumptions of relevance is unproblematic. If one is comparing one Toyota Camry
to another built the same year, and one notes that Camry A starts well in cold
weather, one need make only a few assumptions to judge that Camry B will likely
start well in cold weather—e.g., that they have comparable mileage and wear, that
neither engine has been significantly modified, etc.[11] Many other features of the two
Camrys—color scheme, power or manual windows, leather or fabric interior, etc.—
do not matter when making a comparison between their willingness to start in cold
weather. So, in this case, any reasonable judge easily knows which features of the
case are relevant to the comparison, and which are not. In this case, and in any case
of analogical reasoning, as long as there is (a) a common understanding of what features of a case are relevant, and (b) a large enough base of paradigm/precedent cases
with which to compare current cases (as there is when comparing two automobiles),
casuistic reasoning can work well.

In ethical analysis in a pluralistic society, however, obtaining (a) and (b) can be
troublesome. An appeal to the facts of the case alone cannot suffice to justify a
particular decision regarding what features are the relevant ones and when they are
relevantly similar, since understanding the importance of the facts involves moral
interpretation of them. For example, if a case involves patients in a persistent veg-
etative state (PVS), understanding PVS patients as persons who are already dead
because their upper brains are destroyed will tend to result in different choices of
paradigm case and relevant factors, and thus different recommendations for action,
than will understanding them as patients who are alive but who will die unless they
receive artificial nutrition and hydration.[12] The different understandings of these
patients, which are based on prior differing more general claims about death, are
crucial to knowing what analogies can be made to a case involving PVS.

Without an appeal to principles that operate at a low or intermediate level of generality, one cannot explain what features of a case are relevant; without a shared
understanding of what those principles are, one cannot come to a shared understanding on what features matter, and thus what cases are analogous, and thus what
resolution of the case at hand is justified. Unless a body of shared beliefs can be
derived, or some other means of determining what features are most important, etc.,

---

[10] Sunstein, Cass R. (1993). "On Analogical Reasoning." *Harvard Law Review* 106: 741–91. pp.
746–747.

[11] Example adapted from Sunstein, "On Analogical Reasoning."

[12] See Baruch Brody. (1992). "Special Ethics Issues in the Management of PVS Patients." *Law,
Medicine & Health Care* 20: 104–115.

analogical reasoning cannot really succeed in avoiding the problems of appealing to general principles that are unshared in a pluralistic society. In both selecting the correct paradigm cases, and in reasoning from those cases to the case at hand, casuistry faces difficulties.

A number of casuists actively address the need for common moral guides, and argue that moral thinkers do indeed have access to some more general guides. Jonsen and Toulmin present perhaps the clearest version of this position. They note that there were doctrines of the right and the good throughout all of historical casuistry, and that historical casuists took their role to be working within those doctrines to make those ideals work within the constraints of real life.[13] In their own development of medical casuistry, they also work with some more general claims, or maxims, which are meant to guide one in explaining the case fully in all its detail. These moral "maxims" are simply statements of moral guidelines—e.g., "Don't kill other humans," "Medical treatments should provide expected benefit for the patient," etc.—that are raised by the case. A given maxim will not necessarily motivate the appropriate decision in this case, but they all will help guide the understanding of what the case involves. In their terminology, this is the "morphology" of the case, which "reveals the invariant structure of the particular case, whatever its contingent features. . . ."[14] Then, and only then, one may determine what cases are similar to this case and which are not, with guidance. These "maxims" guide one in determining which features are relevant and which cases, including paradigm cases, are most similar to the instant case.

One can surely note that there are a number of cases in which there is relatively broad agreement as to the right resolution, at least within a given society. Acceptance of the removal of a child from a parent when the parent is abusive or neglectful and rejection of providing unwanted medical treatments to a competent patient who refuses them are two examples of such cases—though determining precisely what constitutes abuse can still remain an open question in some cases. Given the initial agreement on a group of paradigm cases, it seems plausible that the correct maxims can be understood and applied by careful practitioners, and casuists could arrive at solutions to morally problematic cases that could be shared by all those who share agreement on the paradigm cases.

This is the form of casuistry discussed in *The Abuse of Casuistry*, several of Jonsen's articles soon after publication of that work, Strong, and Kuczewski.[15] These

---

[13] Jonsen and Toulmin, *The Abuse of Casuistry*, p. 242.

[14] Jonsen, Albert R. (1991). "Casuistry as Methodology in Clinical Ethics." *Theoretical Medicine* 12: 295–307. p. 301. Strong discusses employing moral principles in much the same way as Jonsen employs maxims. See Strong, "Justification in Ethics."

[15] See, e.g., Jonsen, Albert R. "Of Balloons and Bicycles: The Relationship between Ethical Theory and Practical Judgment."; Jonsen, "Casuistry in Clinical Ethics."; Strong, "Critiques of Casuistry."; Strong, "Justification in Ethics."; Kuczewski, Mark G. (1994). "Casuistry and its Communitarian Critics." *Kennedy Institute of Ethics Journal* 4(2): 99–116; Kuczewski, Mark. (1998). "Casuistry and Principlism: The Convergence of Method in Biomedical Ethics." *Theoretical Medicine and Bioethics* 19: 509–524.

accounts all depend for their function on practical reason, prudence, or some form of Aristotle's concept of *phronesis*, to properly categorize cases. They do not all appeal to maxims per se, but their systems function by utilizing something like a maxim-based approach.[16]

The prudent judge, a person of relevant experience and practical wisdom, analyzes a case, utilizes the more general guidelines to determine which paradigms it most resembles, and determines whether there are any salient differences that mandate deciding this case differently from another case.[17] Based on his wisdom and suitable general guides like maxims, the prudent reasoner is to determine how to resolve the case at hand based on paradigm cases and the resemblances and differences between them and the case at hand. Strong's analysis of Case 4.1 above is an example of this sort of reasoning.

The first concern with appealing to maxims as guides in casuistic reasoning is whether they are sufficient. It is still possible, after a case has been casuistically determined, to ask why case A is most similar to paradigm case B. A casuist may respond to this by noting that a given set of circumstances are similar, and most relevant. But the reasons why those are critical rather than coincidental circumstances in this case need to be clear, as well. This requires maxims; but which, and how? As Jonsen notes, "For most cases of interest, there are several morals, because several maxims seem to conflict. The work of casuistry is to determine *which maxim* should *rule the case* and to what extent."[18] But this means that one needs a way of knowing which maxim should be thought more important in a given case, and maxims alone cannot do this, for one now needs a "higher" level of reasoning to know which maxim ought to rule. If this concern can be resolved, though, there is a more serious problem with guiding casuistry *via* maxims, or the like.

Though casuists do not always tend towards this language, the use of maxims is an appeal towards a "common morality" similar to that to which Beauchamp and Childress appeal.[19] Maxims are thought to be general enough to be shared universally, and consequently can be used as general guidelines to categorize instant cases as similar to particular archetypal cases, as well as enough to guide recognition of relevant features. This is akin to arguing that there is agreement in the common

---

[16] Arras has suggested that in this sort of casuistic reasoning there ought to be explicit rules to explain which circumstances are relevant and why, which would necessarily be theory-laden. These are similar to, but not identical to, maxims. See Arras, John. "Getting Down to Cases: The Revival of Casuistry in Bioethics." p. 40.

[17] No reference to ethical theory is made in nearly all cases. In *The Abuse of Casuistry* Jonsen and Toulmin present an example of a case (ethics of sexuality in the modern age where one can change one's gender) where an appeal to theory is required, because the case is dependent on a new development that changes one of the fundamental features of all prior reasoning (the permanence of biological gender). See pp. 318–322.

[18] Jonsen, "Casuistry as Methodology in Clinical Ethics", p. 298.

[19] As has been noted at least by Beauchamp. See, e.g., Beauchamp Tom L. (2003a). "Methods and Principles in Biomedical Ethics." *Journal of Medical Ethics* 29(5):269–274; Beauchamp, Tom L. (2000). "Reply to Strong on Principlism and Casuistry." *The Journal of Medicine and Philosophy* 25(3): 342–347.

morality on some basic moral guidelines. Whether these guidelines appear more like principles or like moral rules depends upon how one phrases the maxims; but it would not be a stretch to suggest that principles after a few basic specifications could look like the maxims Jonsen and Toulmin present. Insofar as this is similar to a common morality account, it shares some of the same difficulties as a common morality account of principles. The problem with using maxims as guides for analogical reasoning is twofold: first, it is unclear, and unargued for, that the maxims in question are in fact shared by the persons discussing the case; second, insofar as they do exist and are shared, these maxims are like the version of the principles critiqued by Gert, Green, and Clouser. They are sufficiently vague and general that they are insufficient to guide one's decision in a difficult case on their own, as noted by Jonsen in the quote above.[20] Maxims may provide some very basic guidance, but unless made more specific or guided by something more specific, they cannot provide enough to do anything more than begin the analogical reasoning process; and like the principles, once made more specific, they are less likely to be shared. This is a serious problem that keeps maxims, as Jonsen discusses them, from being enough for casuistic reasoning to resolve problems in a pluralistic environment.

Both of these concerns arise in a close examination of Case 4.1. The judgment was made that the relevant features of the case were the patient's faith-based decision to refuse blood products, the prognosis without blood products, the existence of minor children dependent upon the patient for financial and emotional support, and the presence or scarcity of other sources of that support in her absence. Strong argued that these were important features on the grounds that they were relevant to the moral claims "respect the wishes of a competent patient" and "protect the interests of dependent children."[21] It is perhaps fair to assume that these are agreed upon moral maxims, and they do help classify the case. Without something like these maxims, one cannot know that her request and her dependent children are the most important features of the case and not, for example, the gender of the patient, her ability or inability to pay for the procedure, or the non-medical harm to her of a blood transfusion.[22] However, they are not enough to determine the right answer to the case. In order to actually resolve the case, one must also appeal to further general claims regarding the relative comparative importance of these maxims. Such claims are open to rational criticism, which cannot be addressed merely by knowing that "respect the wishes of a competent patient" and "protect the interests of dependent children" are important features of the case. The maxims would need to be specified and/or balanced in a way similar to Beauchamp and Childress's principles, though to be sure this particular language is not always used; thus the same concerns that were presented in the previous chapter will apply to the use of maxims.[23] They are

---

[20] See footnote 18 above.

[21] Actually, in Strong's analysis, he refers to the moral principles of respect for autonomy and beneficence. See Strong, "Justification in Ethics." Either of these function in the same way.

[22] A more in-depth discussion of this latter issue can be found in *Further Troubles*, below.

[23] Jonsen, Albert R. (2000). "Strong on Specification." *Journal of Medicine & Philosophy* 25(3): 348–360. p. 359. Strong, on the other hand, disputes this analysis. Strong, Carson. (2000). "Spec-

useful in categorizing a case in a basic way, but are insufficient to determine what that actual paradigm case should be in a difficult situation.

This can be shown dramatically by the fact that a societal consensus to refuse to force transfusions in cases like Case 4.1 now exists.[24] Strong's analysis may or may not have accurately reflected thought in 1988, when it was published, but it does not now.[25] Assuming Strong was correct about the general consensus at the time, the change in attitudes does not prove Strong's initial casuistic analysis wrong; rather, it shows a change in public sentiment about cases of this sort. But note that the maxims and principles surrounding the case have not changed in the last few decades; they remain "respect the wishes of a competent patient," "protect the interests of dependent children," and the like. Rather, the general societal interpretation of them has changed, and the importance placed on them has changed. If the maxims have not changed but the correct decision has, casuists must be able to explain why.

Note as well that while general public sentiment may have altered, Jehovah's Witnesses have not significantly altered their religious position during that period of time. Those who refused transfusion in 1988 and who refuse transfusion now may well do so for the same reasons, which are not the same as the broadly held view. This will also raise concerns for the practice of casuistry in a pluralistic environment, which is discussed below in *Further Troubles*.

A casuist may address a change in general societal values without a change in general maxims either by embracing the shift in societal values or by providing principled reasons to oppose that shift. One could, in principle, do either, and the maxims and principles guiding the analysis of the case cannot tell one which to attempt. Though maxims provide a more general level of reasoning to assist in guidance of analogical analysis of cases, the proper application of them requires another, more general level of guidance. That the decision in a case can change not because of any facts in the case but merely because of the general opinions of society will create a significant problem for employing casuistry in pluralistic settings—though something like this might enable a more narrow use of casuistry, as elaborated in the final chapter of this work.

The potential for challenge and disagreement with analogical comparisons becomes more problematic the more complex the comparison is; and since many of the difficult cases in medical ethics are indeed quite complex, this can be a significant problem. If one is comparing two similar automobiles, this is rarely a problem.

---

ified Principlism: What is it, and Does it Really Resolve Cases Better than Casuistry?" *Journal of Medicine & Philosophy* 25(3): 323–341.

[24] See, e.g., Hughes DB, Ullery BW, and Barie PS. (2008). "The Contemporary Approach to the Care of Jehovah's Witnesses." *Journal of Trauma-Injury Infection & Critical Care* 65(1): 237–247; Kerridge I, Lowe M, Seldon M, Enno A, and Deveridge S. (1997). "Clinical and Ethical Issues in the Treatment of a Jehovah's Witness with Acute Myeloblastic Leukemia." *Archives of Internal Medicine* 157(15): 1753–1757. This consensus seems to exist in at least North America, Australia, and the UK. Societal consensus, of course, does not nor ever has meant universal agreement. This concern is discussed in *The Existence of Commonly Held Paradigms and Their Limitations*.

[25] Kuczewski, Mark G. "Casuistry and its Communitarian Critics."

But in a more complicated case, more assumptions are needed to make the claim that two cases are comparable: the more features of a case that exist that might be relevant, the more issues that need a judgment as to whether they are relevant, how they are relevant, and whether two cases are therefore analogous. Thus, for a useful moral analogy to be made in difficult cases many more general value claims are required. These claims, in Engelhardt's language, introduce quite a bit of potentially unshared content into any casuistic analysis.

But a reasonable response would be to question the extent to which reasonable challenges might come when at least certain paradigm cases are being applied. Though there can be reasonable disagreement in many cases, perhaps at least some cases are widely accepted as correctly decided, and thus employable as usable paradigms. If so, casuistry would be limited in use in a pluralistic setting, but not helpless, as it would have a potentially significant base of archetype cases with which to work. Morally serious prudent reasoners could then resolve some or perhaps even many cases in medical ethics.

## The Existence of Commonly Held Paradigms and Their Limitations

For some years, there were regular public debates between Tom Beauchamp and Ed Pellegrino about assisted suicide and euthanasia. One point made quite clear by these debates was that they disagree with regard to this issue both in regards to the broad theory as well as practice in individual cases. Either could and should plausibly be called an experienced moral expert in this area,[26] yet they disagree significantly on decisions in cases where persons request assistance in dying. Here— and this is surely not the only situation—two prudent judges disagree. They are aware of the same cases, maxims, and so on, yet they disagree sharply.

Moreover, these two do not only disagree on general issues, such as whether it generally may be beneficent to assist a person in dying,[27] or acceptable to respond to an autonomous request for such assistance, but they also disagree about the analysis of specific cases. For example, Pellegrino is more likely to suspect that a request for death is non-autonomous than is Beauchamp; that is, these two judges may well

---

[26] As this is a rather vague term, proving this point might seem difficult to do. Still, by reasonable measures, both clearly deserve such a title. Beauchamp has edited multiple volumes directly discussing the ethics of assisted suicide or discussing end of life matters including active euthanasia and assisted suicide in depth; the matter has also been a topic discussed in the *Principles* at least since the third edition. Pellegrino is also very well published in this area, with a Medline search of "Pellegrino, ED" and "euthanasia OR assisted suicide" yielding 23 hits, with hundreds of citations of multiple articles.

[27] This point was made clear in many of these debates, and can also be seen in Pellegrino, Edmund D. (1992). "Doctors Must Not Kill," *The Journal of Clinical Ethics* 3(2): 95–102; Pellegrino, Edmund D. (1996). "Distortion of the Healing Relationship", in *Ethical Issues in Death and Dying*, 2nd ed. Tom L. Beauchamp and Robert Veatch, eds. Upper Saddle River, NJ: Prentice Hall, pp. 181–185; and Beauchamp and Childress, *Principles*, 5th ed., pp. 144–148.

perceive the circumstances of a given case differently.[28] These differing interpretations of the circumstances of a case, and which paradigms it most closely resembles,
would motivate different decisions in many cases. Is it most like a case of a reasonable patient autonomously requesting a desired treatment or a frightened person
seeking what falsely seems to be the only way out of an unpleasant situation? They
could conceivably choose different paradigms leading to different conclusions in
the same case; and each could provide what ought to be thought of as good reasons
for their interpretations. Though they would utilize similar maxims and precedential
cases, their interpretations of the facts and the relevant maxims, and consequently
the case resolutions, would very likely differ.

Of course, the casuists who espouse this view are well aware that there are such
disagreements between prudent and reasonable thinkers. They do argue that this
problem ought not be inflated to be more than it really is. Strong notes that, though
Arras is correct to point out[29] that one cannot resolve some major conflicts in medical ethics—such as abortion issues or the appropriate provision of health care to
those who cannot afford it—via casuistic reasoning, this is hardly cause to reject
what casuistry *can* do. There are some issues that are not likely to be resolved by
casuistry, or by most other moral theories in a pluralistic society, either; to reject
casuistry on the grounds that it cannot provide a satisfactory resolution of these
issues is:

> ...plac[ing] unreasonable demands upon it. Casuists neither seek nor expect society-wide
> consensus concerning all conclusions reached by casuistic argumentation. Casuistry could
> be described as seeking conclusions that are hypothetical rather than categorical: they are
> reasonable *if one accepts certain assumptions about ethical values and paradigm cases.*[30]

In the emphasized section, one sees a strong appeal to having something like
a common morality underlying a casuistic analysis. The casuists' position is that
casuistry cannot resolve all moral conflicts, but insofar as we have some common
agreement, casuistry will be able to resolve many. Casuists also argue that there
is a large quantity of agreed upon moral claims, especially about paradigm cases,
in society in general. As noted earlier, this fact was part of the motivation for *The
Abuse of Casuistry*, since Jonsen and Toulmin noted that the widely disparate group
that made up the National Commission for the Protection of Human Subjects of
Biomedical and Behavioral Research was able to agree on the appropriate resolution of many, though not all, particular cases that they considered, even though their
theoretical interests and beliefs varied dramatically.[31] The argument could be made
that there is a widespread social agreement on the correct resolution of a wide variety of paradigm cases that can serve as the basis of a casuistic reasoning; persons
need not share agreements on moral theory to have the basis for casuistry to function
to resolve problems. There is widespread agreement that American and Nazi cases

---

[28] See footnote 27.

[29] Arras, John. "Getting Down to Cases: The Revival of Casuistry in Bioethics."

[30] Strong, "Critiques of Casuistry", p. 405. Italics changed from original.

[31] Jonsen and Toulmin, *The Abuse of Casuistry*, pp. 16–19.

of experimentation on humans without their consent were wrong, that competent persons may refuse even life-saving medical care, that needed life-saving medical care may be imposed on an infant even if the parents object, as in Case 2.2 in Chapter 2; the list could go on. Casuistry will not be able to resolve all problems, in particular those where there is no shared agreement on paradigm cases or their basic descriptions. Abortion cases and whether they ought to be classified as cases of preserving bodily integrity, or as cases of killing another, or as something else entirely, will remain troublesome. But there will still be a significant space for and value of casuistry because there is this reasonably large base of agreement on paradigm cases.

However, it is also important not to overestimate the importance of this shared agreement. Even where there is agreement on a basic paradigm case, the extent to which that agreement can yield agreement on any related cases can be severely limited. Consider a case presented by David DeGrazia, which he takes from *Crime and Punishment*:

> With the cry of "now," the mare tugged with all her might, but far from galloping, could scarcely move forward; she struggled with her legs, gasping and shrinking from the blows of the three whips which showered upon her like hail. The laughter in the cart and in the crowd was redoubled, but Mikolka flew into a rage and furiously thrashed the mare, as though he supposed she really could gallop.
>
> "Let me get in, too, mates," shouted a young man in the crowd whose appetite was aroused.
>
> "Get in, all get in," cried Mikolka, "she will draw you all. I'll beat her to death!" And he thrashed and thrashed at the mare, beside himself with fury...
>
> All at once laughter broke into a roar and covered everything: the mare, roused by the shower of blows, began feebly kicking....
>
> Two lads in the crowd snatched up whips and ran to the mare to beat her about the ribs....
>
> ... He ran beside the mare, ran in front of her, saw her being whipped across the eyes, right in the eyes!... She was almost at the last gasp, but began kicking once more.
>
> "I'll teach you to kick," Mikolka shouted ferociously. He threw down the whip, bent forward and picked up from the bottom of the cart a long, thick shaft, he took hold of one end with both hands and with an effort brandished it over the mare....
>
> And Mikolka swung the shaft a second time and it fell a second time on the spine of the luckless mare. She sank back on her haunches, but lurched forward and tugged forward with all her force, tugged first on one side and then on the other, trying to move the cart. But the six whips were attacking her in all directions, and the shaft was raised again and fell upon her a third time, and then a fourth, with heavy measured blows....
>
> "I'll show you! Stand off," Mikolka screamed frantically; he threw down the shaft, stepped down in the cart and picked up an iron crowbar. "Look out," he shouted, and with all his might he dealt a stunning blow at the poor mare. The blow fell; the mare staggered, sank back, tried to pull, but the bar fell again with a swinging blow on her back and she fell on the ground like a log.
>
> "Finish her off," shouted Mikolka and he leapt, beside himself, out of the cart. Several young men, also flushed with drink, seized anything they could come across-whips, sticks,

poles-and ran to the dying mare. Mikolka stood on one side and began dealing random blows with the crowbar. The mare stretched out her head, drew a long breath, and died.[32]

DeGrazia assumes, hopefully correctly, that there is a widespread agreement that the actions of Mikolka and the other men in this case were unacceptable. This could be a paradigm case of the wrongness of harming animals, and as such could form the basis for a casuistic argument regarding other cases where animals are harmed. So, perhaps this paradigm case will assist in addressing the head injury experiments at the University of Pennsylvania made (in)famous by the video *Unnecessary Fuss*, made from videotapes stolen from the lab by members of the Animal Liberation Front:

> Baboons were deliberately given severe head injuries by means of a hydraulic piston attached to a helmet that was placed over the baboon's head. The piston and helmet device would jerk the baboon's head to the side at a force of up to two thousand times the force of gravity, causing serious injury by the impact of the soft brain tissue against the inside of the skull. The damage was severe and often resulted in partial to total paralysis or coma. Because tests of motor reflexes before and after injury were being done, anesthesia was not provided, and the animals were conscious through the entire procedure. The baboons were then kept alive and studied over the course of a few months, sometimes receiving additional head injuries during this time, after which they were killed and their brains were examined.[33]

While much of the official debate about this case revolved around whether the appropriate laws and regulations had been followed in performing these experiments, surely at least a part of the moral question involved is whether or not it is permissible to so harm these animals, even if regulations were followed. But even though the case of Mikolka and the mare is an excellent paradigm case of the wrongness of harming animals, the agreement on that paradigm case will not carry far into the debate of whether scientists may perform head injury experiments on baboons, or indeed little else about treatment of animals. For while there may be agreement on it being wrong to harm animals in Mikolka's case, there is little agreement as to what features about the case are important, what parallels to the head injury case exist, and which of those matter, and how much. Some persons will judge that the expected benefits to humans from the U. Penn experiments were great enough to make the harms to animals unfortunate, but acceptable; some will argue that any expected benefit to humans is sufficient to make the harms acceptable. On the latter account, all that was wrong in Mikolka's case was that he was harming the mare for no reason. Some might argue that the suffering of the baboons is sufficiently different, perhaps because it was less painful, than that of the mare to make the two cases significantly different. Others will argue on the other side that the expected

---

[32] Dostoevski, Fyodor. (1982). *Crime and Punishment*, trans. Constance Garnett. New York: Bantam Classics, pp. 50–52, quoted in DeGrazia, David. (1996). *Taking Animals Seriously*. New York: Cambridge University Press, pp. 40–41.

[33] Case adapted from Orlans, F. Barbara, Tom L. Beauchamp, Rebecca Dresser, David B. Morton, and John P. Gluck. (1998). *The Human Use of Animals*. New York: Oxford University Press, pp. 71–88.

benefits were not sufficient to justify the experiments; others will also argue that no expected benefits could justify this kind of harm, especially to primates; still others will argue that it is wrong to intentionally harm any animals even if there is a significant expected benefit. The agreement that society had on the paradigm case rapidly withers to irrelevance when a troublesome case rears its head; and notice that in each of these various interpretations different circumstances about Mikolka's case are often taken to be relevant—e.g., whether a benefit arises from the harm, whether the animal is a "higher" animal, or even a primate, or not, etc. Prudent judges in this case are split among the various resolutions as well. Without further guidance from more general values or principles that do not seem to be shared, no conclusions can be drawn about more difficult cases from this paradigm case. The paradigm is indeed held widely; virtually no conclusions in more difficult cases can be drawn from that without more theory to explain what features about it are important.

## Further Troubles: The Problem of Hidden Assumptions and Values

Moreover, the extent to which there is shared agreement on paradigm cases and their resolutions may be far less than casuistry's proponents argue. In fact, much of the "shared agreement" that may exist is based on assumptions which, when revealed, can be seen to be unshared in a pluralistic society. This can be seen by expanding on an argument made by Tomlinson.[34]

Tomlinson rightly argues that Strong's resolution of the Jehovah's Witness case above is guided by the particular choice of paradigms, but would plausibly be different if different paradigms were chosen. He suggests that another paradigm that fits the case well would be the removal of a kidney from a parent, against her will, in order to donate it to a child who needs it to get off dialysis. This, too, is a patient who has refused a medical action, the implementation of which would greatly benefit her child. Various courts have decided that this sort of action cannot be imposed upon the parent. The medical ethical community is at best divided on such cases, where most would argue that this is a case where one cannot and ought not force the operation. Therefore, Tomlinson argues, "[I]sn't forced organ donation a paradigm case of a competing type that supports a decision *not* to transfuse her against her will, even for the sake of her children's health?"[35] He argues that one cannot show that the above Case 4.1 ought to be resolved by obtaining a court order to transfuse this patient unless one can show that the paradigm of forced organ donation is the wrong one to apply. Arguments of this sort would defend the general conclusion now accepted that we ought not transfuse.

This claim is true enough, but there is another point which can be made regarding this, which is that Strong, and many others who deal with cases involving Jehovah's

---

[34] Tomlinson, "Casuistry in Medical Ethics: Rehabilitated, or Repeat Offender?", pp. 13–19. He makes a different point with this example than is made here.

[35] Tomlinson, "Casuistry in Medical Ethics: Rehabilitated, or Repeat Offender?", p. 13.

Witnesses, seem to use a particular set of underlying and unshared value claims even to frame the case as Strong has done. Once one realizes the existence of these value claims, the depth to which casuistry is limited by the analogical reasoning it depends upon can be shown.

It is true that a blood transfusion carries very little *medical* risk to its donor or recipient, while an operation to remove a kidney is significantly more risky; thus, Strong could argue, this factor is one that makes these two cases relevantly different. But it is a feature of these cases involving Jehovah's Witnesses that the patients in them believe very strongly that a great harm will be done to them by forced transfusion. It may lead to forced expulsion from the Witness community; depending upon the patient's particular interpretation of her faith, it may even prevent her from achieving happiness in the afterlife. This issue is never raised in Strong's analysis of the cases, and that places significant doubt on the rectitude of forcing a transfusion in any of the three cases. That is, Strong never takes seriously the issue of harm to the patient. Though he rightly notes that there is an issue of respect for the patient's autonomy in each case, the claim that there might be significant harm to the patient is not considered in any of the cases. If one does take this claim seriously, the rightness of forcing transfusion in the presumed paradigm case of Case 4.2—and thus, of course, the rightness of forcing transfusion in Case 4.1, as well— is placed into serious doubt.[36] Tomlinson's paradigm case of forced kidney donation may be a good paradigm for forced transfusion to a Jehovah's Witness precisely because it takes seriously the harm understood by the patient, putting it into a context more comprehensible to non-Jehovah's Witnesses. In fact, it still rather radically *under*states the harm of the transfusion to a strict Jehovah's Witness: the operation to remove a kidney presents a real but small risk of loss of life or health, while a forced blood transfusion presents a risk ranging from minimal to the loss of eternal happiness. Since society has not come to an agreement that one may be forced to risk the harm of an organ donation against one's will, it seems reasonable to think that a case in which a person is forced to risk something even greater could not possibly be a paradigm case where one would be required to run that risk.

Yet no notice is made of this particular harm in the above cases. To ignore a harm, or significant risk of harm, to a patient normally would make an ethical analysis severely lacking, yet little mention is made of this concern in Strong or the critiques of his position. The only valid reason for ignoring a possible harm to a patient would be that there is, in fact, no actual risk—or at most a negligible risk—of harm to the patient. (For example, one would be justified in ignoring the harm from satellite-mounted mind-control beams risked by a patient by removing his tinfoil-lined hat in order to treat a serious scalp laceration.) Perhaps the reason that Strong, and society in general, do not treat transfusion as having similar or worse risks of

---

[36] This, of course, means that Case 4.2 really cannot be a paradigm case, as the resolution of a paradigm case is supposed to be clear to virtually all prudent judges. To avoid confusion, the language will be retained here.

harm as a kidney donation, or really as having any significant harms at all, is that most non-Witnesses do not actually believe there is such a harm. The decision made in Case 4.1 therefore depends on a basic principle regarding the potential harms of the treatment that is explicitly not shared by the patient. The fact that society no longer accepts Strong's initial interpretation does not change this. Even the respondent who argues that we cannot require transfusion because we must allow people the autonomous right to make medical decisions for themselves, even at risk of harm to themselves or even loss to others, fails to take the potential harm of blood transfusion seriously. Essentially, this claim is, "People have the right to make foolish decisions." Yet, if one takes the Jehovah's Witness claims seriously, the decision to avoid transfusion is anything but foolish. When one states, "A blood transfusion carries with it significant (medical) benefit and virtually no risk of (medical) harm," that is a true statement for the Jehovah's Witness only if one includes the parenthetical adjectives. It is a significant source of non-medical harm; and Jehovah's Witness patients are often quite reasonably far less concerned about the medical harms than they are of these other spiritual and social harms. There is an implicit and hidden assumption, necessary to make the resolution of the case that Strong does, that a blood transfusion carries no risk of harm except the minimal medical one.

This alone may make it clear that neither Strong's resolution nor the updated reversal of it may be acceptable in a pluralistic society, in which the value claim that underlies both decisions is not universally shared. Such a claim is not defensible in a pluralistic society. One can only claim that transfusions provide no harm if (a) one does not actually take seriously the faith-based beliefs of Jehovah's Witnesses, or (b) in the casuistry of medical ethics one is concerned only with the strictly medical effects of a particular action. But neither of these is an acceptable response. The first is problematic in a pluralistic society: respecting persons' autonomy by letting them choose only versions of the good life that the majority generally accepts as valid is not a satisfactory approach to the problem of pluralism. In any case, it is not immediately clear on what grounds a casuist could defend this claim, and it is less clear why a Jehovah's Witness patient ought to accept such reasoning as justified. The second response, though perhaps initially more promising, fails as well because it hamstrings the very argument it is meant to support. Case 4.1 was thought most similar to Case 4.2 because of the concern for the patient's surviving children, which is a concern not directly related to the medical effects of the treatment. If one is concerned only with the medical effects of an action, this is not a morally difficult case, as the harm to the children is not medically related to the therapy in question.

It is clear that a decision-maker who was also a Jehovah's Witness would take the harm of the transfusion far more seriously, resulting in a very different analysis of the case. Different reasonable persons can, and in many cases will, make different judgments of the same case because their basic views of what is important and what is less so may differ and should inform their decisions. Note how similar this is to the concerns raised with the common morality view in the previous chapter. Because casuistry seeks to function from cases upward, yet actually depends on a more general level of judgments to decide those cases, it cannot be separated from the moral and cultural beliefs of the decision-maker; yet in a pluralistic society those are not shared.

## A Possible Response: Medicine Provides the Maxims

Perhaps there is a claim set that can be used to ground modern medical casuistry. Consider a different sort of challenge to Strong's resolution of Case 4.1. In this case, a new paradigm case is presented that represents a different option than any suggested by the other paradigms presented by Strong or Tomlinson. This fourth possible paradigm case may provide a better resolution of Case 4.1 than either of the two possible solutions suggested by Strong, or the third presented by Tomlinson. An argument to defend this solution is less important than noting what a casuist's inability to address this solution means regarding what casuistry can and does address.

> **Possible Paradigm Case 4.4:** A 35-year-old Jehovah's Witness woman arrives in the emergency room with significant blood loss after a serious automobile accident. A blood transfusion is determined to be necessary for her survival, but she refuses all blood products. Adequate substitutes are not readily available, and doctors are certain she will die without a transfusion. Her husband and family support her in this decision, despite the fact that she is the chief breadwinner in the family, which includes several young children. No other extended family members can be found that can help the family through this. The chief resident on duty in the ER happens to be the next-door neighbor of the family, and insists that her views be respected. When the needs of the family are raised to him as a potential moral complicating factor, he assures the remainder of the staff that he and his family will provide financial and social support to the family as much as necessary. No transfusion is provided and the woman dies.

In this case, another option beside transfusion or harm to the patient's children is presented. A court order to transfuse the patient is not sought, but rather a means of providing financial and social support to the family and children in question is sought. This has the immediately clear value of promoting both of the moral issues thought important at the beginning of the case. It respects the authority of an autonomous person to choose how she is treated, and it also positively addresses the harms to others the death of the patient would cause. It does not explicitly address the harm of transfusion to the patient, but it consistently could. Moreover, it may be that Case 4.4 is relevantly similar enough to Case 4.1 that this should be the appropriate paradigm. It is not likely that the chief resident in Case 4.1 is the next-door neighbor to the family in question, but the husband is a worker at a private club, where perhaps someone else might be able to provide the support needed. Possibly the church community could provide some support, or the family's actual neighbors; there are other possibilities as well. In all of these ways it may be so that Case 4.1 is closest in its circumstances to Case 4.4, in which case the transfusion should not be forced on the patient.

What is interesting is not whether Case 4.1 actually is more similar to Case 4.2 or Case 4.4—though that does seem important to know when resolving the case—but rather that, in Strong's initial presentation of the case, and in the later analysis of it both by Strong and others, we do not have the information necessary to know whether Case 4.1 is more like Case 4.2 or Case 4.4. In the description of the case, the various social connections of the patient and her family were not even known, or in any case not considered important enough to be discussed. In any actual case, of course, there is a virtual infinitude of possible facts that can be listed about it; in choosing the facts that might be important out of that unmanageable list one again must use some sort of criteria for sorting the more from less important facts. The facts that are given, and those that are not, give strong clues to the thought process used to parse the facts. A physician's case description would include many medically relevant features, such as a careful description of diagnosis and prognosis, including such features that are probably irrelevant to an ethical analysis, such as gender and exact age, but which are often important in medical diagnosis. A social worker writing up the case would probably pay less attention to medical facts and focus instead on the interactions between family members and other related persons or individuals otherwise close to the family. The decision as to which factors are thought important to consider is already determined by one's mindset.

The important feature of this is not the potential problem of limitations of the medical view nor questions of the possible arbitrariness of choosing some facts over others, but rather that, when one practices casuistry in a medical environment, one is practicing casuistry in a particular setting, with its own particular views of what is more important and what is less so. Based on this recognition, one could argue that casuists doing medical ethics should practice casuistry in the context of the worldview of medicine. In other words, one may recognize that casuistry can only be performed in the context of a particular set of more general views, and that there are multiple possible sets of such views; yet one may argue that priority in this case should go to the set of views intrinsic to the practice of medicine itself.

Recall Strong's claim about what casuistry could be perceived as doing, quoted above: "[c]asuistry could be described as seeking conclusions that are hypothetical rather than categorical: they are reasonable *if one accepts certain assumptions about ethical values and paradigm cases*."[37] What assumptions about ethical values and cases could one assume in a modern pluralistic society? Rather than looking to the common morality, which has the problems enumerated above, one can look to the practice of medicine for some basic values and paradigm cases. That is to say, a bioethical casuistry is a casuistry *of medicine*, and one can and should practice casuistry of medicine in the context of the basic values and paradigm cases inherent in that practice.

This position is explicitly held by Tallmon, and is highlighted in Jonsen's later writings where he more strongly elaborates the importance of the special topics of

---

[37] Strong, "Critiques of Casuistry", p. 405. Italics changed from original.

medicine.[38] The existence of an internal morality of medicine is also argued for, not directly in connection with casuistry but not in a fashion inimical to it either, by Miller and Brody[39] and in a different fashion by Pellegrino.[40] The focus on the morality inherent to medicine allows one to avoid the wide pluralism in the larger context of society, while giving one some specific features, goals and guidelines that are specific to and constitutive of the practice of medicine and thus are consistent in that practice. A "morality of medicine" that describes the various values and can include commitments to particular principles, rules, and cases necessary for the practice of modern medicine would provide sufficient higher-level theory for the practice of a casuistry of modern medicine.

One may find this internal morality by exploring first the basic function of medicine, as Miller and Brody seek to do by a quote from Dan Callahan:

> (i) 'the prevention of disease and injury and promotion and maintenance of health'; (ii) 'the relief of pain and suffering caused by maladies'; (iii) 'the care and cure of those with a malady, and the care of those who cannot be cured'; and (iv) 'the avoidance of premature death and the pursuit of a peaceful death.'[41]

This examination can further expand to an examination of medical codes of ethics and codes of professional standards for general moral principles and values, taking care that one does not include contested values while calling them uncontested. One could also look to the body of cases that are maintained as a part of the medical moral understanding, which gives one the key cases and paradigms that will be a part of the taxonomy of cases in medical casuistry.[42] Classic cases such as those of

---

[38] Tallmon, James M. (1994). "How Jonsen Really Views Casuistry: A Note on the Abuse of Father Wildes." *Journal of Medicine and Philosophy* 19: 103–113; Jonsen, "Casuistry: An Alternative or Complement to the Principles"; Jonsen, Albert R, Mark Siegler, and William Winslade. (1992). *Clinical Ethics*, 3rd ed. New York: McGraw Hill.

[39] Miller, Franklin G. and Brody, Howard. (2001). "The Internal Morality of Medicine: An Evolutionary Perspective." *Journal of Medicine and Philosophy* 26(6): 581–599; Miller, Franklin G. (1998). "The Internal Morality of Medicine: Explication and Application to Managed Care." *Journal of Medicine and Philosophy* 23(4): 384–410. They do not claim to be providing sufficient grounds for the whole of medical ethics, but what they do provide are guidelines that could be useful in classifying cases and could serve as casuistic maxims.

[40] Discussed recently in Pellegrino, Edmund D. (2001). "The Internal Morality of Clinical Medicine: A Paradigm for the Ethics of the Helping and Healing Professions." *Journal of Medicine and Philosophy* 26(6): 559–579. This is not to suggest that either Pellegrino or Brody and Miller would agree with the analysis put forth herein; but they have developed useful beginnings for this thought process.

[41] Miller and Brody, "The Internal Morality of Medicine: An Evolutionary View." p. 582, quoting Callahan, D. (1996). "The Goals of Medicine: Setting New Priorities." *Hastings Center Report*, 25: S1–S26.

[42] Pellegrino would likely object to this as including too much social construction into the morality internal to medicine. See Pellegrino, "The Internal Morality of Clinical Medicine," pp. 563–565.

Karen Ann Quinlan,[43] Tatiana Tarasoff,[44] and Timothy Quill and "Diane",[45] as well as cautionary cases such as *Buck v. Bell*[46] and the case of Dax Cowart[47] will also be a part of the medical community's lexicon, as may be many other cases in literature or in remembered history.

One could also look to the practices of modern physicians and the thinking and writing that comes from this. By doing this, one will discover rules and principles that have been and continue to be a part of medical practice, and which could serve as maxims and guidance for maxims. Basic Hippocratic ideals such as "Help, or at least do no harm," and "Act as an appropriate caretaker of your patient," will be a part of the practice or "community" of medicine, as will more recent ideals such as the practice of informed consent. This process is also continual, as "[d]octors, nurses, scholars, lawyers, administrators, and legislators are engaged continually in a dialectic regarding medical standards"[48] and judging practices, rules, and cases in such a way to continually develop and maintain a basis for a morality of the practice of medicine. This process will not derive as extensive a moral system as the medieval Catholic morality that undergirded medieval theological casuistry, but there may well be enough of a set of accepted practices, rules, principles and cases to support casuistry in the practice of medicine—which, after all, is not meant to be as extensive as the medieval casuistry that was meant to encompass the whole of the moral life.

Though one must appeal to theoretically unshared norms and values to understand what paradigms and analogies are appropriate, on this account those norms may be found in this "morality of medicine" and are inherent in the practice of medicine. Thus, they ought to be shared by anyone involved in medicine, at least implicitly. The shared set of moral claims is more limited than the field of all things medical—for example, it does not and cannot encompass such things as the overall rightness or wrongness of abortion, especially as a matter of public policy—but, the argument would go, it can encompass enough norms, cases and judgments to make casuistry very useful in medicine.

---

[43] See Supreme Court of New Jersey, "In the Matter of Karen Quinlan, An Alleged Incompetent", in *Ethical Issues in Death and Dying*, 1st ed., Robert F. Weir, ed. New York: Columbia University Press, 1977, pp. 274–7.

[44] California Supreme Court. *Tarasoff v. Regents of the University of California*. 131 California Reporter 14. Decided July 1, 1976.

[45] Quill, Timothy E. (1996). "Death and Dignity: A Case of Individualized Decision Making", in *Ethical Issues in Death and Dying*, 2nd ed. Tom L. Beauchamp and Robert M. Veatch, eds. Upper Saddle River, NJ: Prentice Hall, pp. 156–160.

[46] *United States [Supreme Court] Reports 274*. (1927), 1000–1002. Reprinted in abridged form in *Contemporary Issues in Bioethics*, 4th ed. (1996). Tom L. Beauchamp and LeRoy Walters, eds. Belmont, CA: Wadsworth Publishing Co., pp. 607–608.

[47] Reports in various forms are available, including (1998). "A Demand to Die", in *Cases in Bioethics: Selections from the Hastings Center Report*, 3rd ed. Bette–Jane Crigger, ed. New York: St. Martin's Press, pp. 110–112.

[48] Tallmon, "How Jonsen Really Views Casuistry", p. 111.

The development of the content of that community, and the casuistic analyses that would follow from it, would take a lot of work, which has not even been begun here. The very possibility of such a morality of medicine existing in an era of deep disagreements on moral views has also been questioned, which this work will also not directly address.[49] Of interest here is an examination of the strengths and limitations of such a possible morality, and casuistry based on it, assuming that such an internal morality could be derived.

The first concern with a morality of medicine is that it must tread the same impossibly fine line that some hoped principles to be able to traverse, as discussed in the previous chapter. The medical morality may either provide relatively uncontroversial, but unusefully vague rules, analyses of cases, and so on, or it must take a stand in particular cases of disagreement and choose one way or another. Given that part of the development of the medical morality comes from court cases and legislative policies, which make definitive judgments and guidelines in such cases, it is possible that it could be understood as choosing the latter. But what this means is that the "medical morality," defined as those who agree with the moral direction and decisions of society, will be composed of only some, and possibly even not very many, of the persons involved in the provision and receipt of medical care. Jehovah's Witnesses are concerned with preserving their lives, which is a value found in the medical morality, but not at the cost of accepting blood transfusions, which the medical morality would reject. Roman Catholics are concerned with preserving the lives of humans in a fetal state as well, and at least some are similarly concerned with human embryos. Singerians are interested in the welfare of non-humans as well as most humans, though not necessarily embryonic or upper-brain-dead humans. At this point in time, the practice of medicine conflicts with most or all of these viewpoints—though there has been extensive recent research into treatment of Jehovah's Witnesses, including bloodless alternatives,[50] court orders for blood transfusions are often requested for underage persons and even some competent adults. Abortions are performed in most Western countries, as of this writing, embryonic stem cell research in the United States is significantly restricted only by a hotly contested executive order and two Presidential vetoes,[51] and animal experimentation is the norm in medical research. Either the persons who reject these actions are not a part of the "medical morality," in which case it is not clear why they ought to care what the moral decisions made by casuistry within the medical community are, or the medical morality contains disagreement and dissent, in which case casuistry

---

[49] Veatch, Robert M. (2001). "The Impossibility of a Morality Internal to Medicine." *Journal of Medicine and Philosophy* 26: 621–642; Wildes, Kevin W. (1994). "Respondeo: Method and Content in Casuistry." *Journal of Medicine and Philosophy* 19: 115–119.

[50] Nearly 200 articles on Jehovah's Witnesses appear on a Medline search covering the years 2001–2006. Many of these are case studies involving use of bloodless alternatives for medical procedures; many others develop ethical analyses of cases where medically necessary blood products are refused.

[51] As of printing, this has been at least partially reversed by order of the new Presidential administration.

based on it will not necessarily lead to a clear decision in a given case. Even if one develops a medical morality that does lead to a clear decision, there can be persons in the medical community who will reasonably deny the validity of those decisions. Even the previously determined basic principles of medicine—"Informed consent is obligatory for treatment of an autonomous patient," "Medical treatments ought to benefit the patient," and so on—cannot define a unified set of medical practices that all, or even most, persons will accept because of different understandings of what benefit is, when harm occurs, who the patient should be understood to be, and in what extraordinary cases these rules can be set aside. These are the issues that need to be shared to obtain broad agreement on a particular analogous case; but they are not.

So if a single "medical morality" is derived, which is what the medical community, as described above by Tallmon, seems to be trying to do through court decisions, etc., it is unclear what importance it ought to have for any individual. For those who rationally disagree with the relevant moral claims of the medical morality, including the Jehovah's Witness patient in Case 4.1 above, there seems to be no more reason to accept the moral validity of the conclusions of the casuistry of the medical community than there would have been for a medieval Jew to have accepted the moral rectitude of a decision of a Jesuit confessor regarding the Jew's actions. But if this is true, then the entire basis for employing casuistry in a pluralistic environment seems to have vanished, since the practice can provide no morally compelling reasons for choosing a particular course of action over another to persons who disagree.

This does not mean that analogical reasoning is irrational, of course, or even inappropriate in many of the places it is used. It is, as has been argued,[52] well suited for reasoning within the law, where there is a base of shared claims (precedent) in the foundational documents of the society and the common law. It is also probably true, as Wildes has argued, that reasoning by analogy is very well suited for case-based reasoning within a moral tradition such as Roman Catholicism.[53] But neither of these can represent moral reasoning in a pluralistic society, as the former involves a strict hierarchy of legal authority and a reasonably extensive common legal world view, and the latter involves a common moral world view and recognition of "communal structures of moral authority and interpretation."[54] There could be multiple medical moralities premised upon basic belief sets. Roman Catholicism would have its own medical morality, Singerians would have their own, and so on.

For obvious reasons, this approach will not serve to allow casuistry to resolve moral conflicts in a pluralistic environment. If a variety of different moral medical communities were derived, each by the different sets of persons who hold different views, the results would be odd. The Jehovah's Witness patient would be in one

---

[52] Sunstein, "On Analogical Reasoning."

[53] See Wildes, Kevin Wm. (1993). "The Priesthood of Bioethics and the Return of Casuistry." *Journal of Medicine and Philosophy* 18: 33–49.

[54] Ibid., p. 35.

medical moral community while her secular humanist doctor would be in another; this means, again, that the decisions of these persons following differing medical moralities would have no particular moral "pull" on each other. Their decisions and decision-making processes would be structurally similar, but the content would differ; and without that shared content, there would be no reason for each one to think the other had correctly resolved the case. Something like this approach may be helpful however, which will be explored in the next chapter.

Therefore, casuistry alone cannot resolve moral conflicts in medicine in a pluralistic society because reasoning by analogy cannot justify the resolutions of individual cases without an appeal to more general shared norms and values which cannot be justified across a pluralistic society. Casuistry can function within a moral community to resolve problems, because in a moral community there are such agreed upon norms and values, but absent those in a pluralistic society, the method cannot reach or justify its desired conclusions. Casuistry cannot resolve the problems found in medical ethics in a pluralistic environment.

# Chapter 5
# Moral Acquaintanceships as a Means of Conflict Resolution

The preceding chapters have discussed three different attempts to determine and justify actions to be taken to resolve potentially morally problematic cases. Recall from Chapter 1 that the definition of "resolving" difficult cases is presenting and defending a particular action as the right one to take in a potentially morally difficult circumstance, and justifying that decision as correct to others in a way that those others can (or should be able to) recognize as a valid justification. Note that this does not require that each person actually agree that each other person's justification is adequate to justify the action to themselves, but each person must be able to recognize the justification as a valid form of justification, one which could possibly serve as a justification for the decision for another person with a different moral understanding. A reasonable justification, then, is a justification that is at least a well-formed justification, grounded in moral claims, which one recognizes as a valid form of moral justification.

This requirement for public justification is weak, in that it requires only that one accepts that the justifications proffered by others for this decision are plausible justifications, not that one accepts that they are correct. Yet it is also a demanding requirement in that it requires that each person actually hold that the action in question is the right action to take in the given case, and is actually justified as the right action to take by good reasons, and not merely that it is the best of a bad set of available choices or the best compromise position available.[1]

---

[1] This does not imply that "Action X is the best compromise" may not sometimes be a part of a valid justification for an action, but it is not always part of a valid justification. There are also other reasonable definitions of resolution, not appealed to here. For a discussion of various meanings of resolution, see Beauchamp, Tom L. (1987). "Ethical Theory and the Problem of Closure", in *Scientific Controversies: Case Studies in the Resolution and Closure of Disputes in Science and Technology*, H. Tristram Engelhardt, Jr. and Arthur L. Caplan, eds. New York: Cambridge University Press, pp. 27–48. For example, one might achieve closure in a given case (and by that resolve the case in a sense) if one can determine a compromise position that, though no one necessarily thinks is the best action to take, is one to which all will accede as an adequate compromise. This could be attained by what Beauchamp calls "negotiation closure," defined as settling a controversy through an intentionally arranged and morally unobjectionable resolution acceptable to the principals in the controversy. It is reasonable that the theories discussed herein may be able to resolve many moral problems in this sense of the term. Conversely, there are more demanding definitions

S.S. Hanson, *Moral Acquaintances and Moral Decisions*, Philosophy and Medicine 103, DOI 10.1007/978-90-481-2508-1_5, © Springer Science+Business Media B.V. 2009

Because resolving a difficult moral case usually requires justifying the chosen action to more than one person, some form of group resolution is required in order to adequately resolve a difficult case. Though the definition of resolution in this work could function as an individual form of resolution—e.g., a moral case is resolved to that individual if, and only if, that individual holds a given decision to be the right decision and has an adequate justification for holding that claim—it also functions equally well as a form of group resolution. As a group resolution, a case can be considered resolved to a given group if, and only if, all persons in that given group of persons hold that a given decision is the right one and that the justifications proffered for that decision by all persons in that group are valid forms of justification for that decision. One should therefore note that when a case is resolved on this account, it is resolved in the context of a particular group of persons; and it is only necessarily considered resolved to that particular group of persons. It therefore follows that a decision could be resolved on this account to one set of persons but not to another set of persons, because one or more members of the latter set either rationally disagree that the decision is the right one or rationally disagree that the justification given by one or more of the others in the group is an acceptable justification.

On this account of resolution, Engelhardtian theory can successfully resolve only the few claims that Chapter 2 argues can be defended through the "principle of reason-giving," and it fails to be able to actually determine right actions in most of the cases that it seeks to be able to address. It definitely cannot resolve the sorts of cases one needs to be able to address in medical ethics in a pluralistic society. Though one could justify some extremely minimal claims to reasonable moral persons as Engelhardt seeks to, one cannot justify Engelhardt's own principle of permission, which is the means he uses to make his theory useful for addressing difficult cases, on the grounds he sets.

Principle-based theory provides a language for persons to analyze and discuss morally problematic cases, even if they are, in Engelhardt's language, moral strangers; casuistry seeks to find paradigmatic cases that can be extended analogically to assist in resolving less simple cases. Both fail to be sufficient to resolve for the whole of society many of the problems that one encounters in modern medicine in a pluralistic society.

If all that can be done in a content-free fashion is to employ the principle of reason-giving, then secular bioethics is in serious trouble. Very little indeed could be done with a secular bioethics based on that principle alone, as noted in *Back to Nihilism?* in Chapter 2. Given the arguments against the content-free nature of the principle of permission, and given the arguments in Chapters 3 and 4 that neither principle-based theories nor casuistry will be able to present a minimal amount of

---

of resolution: a resolution where all parties agree upon a single complete moral argument and conclusion (what Beauchamp calls "sound argument closure") is a more demanding version, which cannot be attained in all cases. There are thus a variety of definitions of the term "resolution" which could be adequate definitions; they are not the definition used here.

content that necessarily should be agreed upon by moral strangers, it seems that secular bioethics may have only the principle of reason-giving to which to appeal. If this is the case, then secular bioethics is not much better off than nihilism.

This is not to say that these theories can accomplish nothing. As argued below, these methods can be successful in addressing and resolving some morally problematic cases, as long as those cases are in particular contexts. As was seen in Chapters 3 and 4, it is possible for principles and/or casuistry to serve as a means for persons to resolve a difficult moral case within the context of a moral community; in this chapter I will show that this methodology can be extended to resolve moral cases within groups of persons with rather disparate moral views.

The most important result from the discussions of principles and casuistry does not regard any inability to resolve morally problematic cases, but rather an understanding of their limitations in resolving them. Appeals to specified principles in the context of a coherent system, or appeals to analogous cases and appropriate understandings of what is analogous, can address and even resolve cases in certain circumstances. It is not the failures of principle-based theories and casuistry that are most interesting but their (limited) successes, and the contexts in which those successes can occur. An exploration of these successes, in order to understand how they can work and if, and in what contexts, moral resolutions to morally problematic cases can be derived, is the clear next step in order to avoid the near-nihilism of the secular ethics of reason-giving.

## The (Limited) Usefulness of Principle-Based Theory

Principle-based theory seeks to serve as a means for different persons to find a common language by which to address and hopefully resolve morally problematic cases. It means to do this by providing a framework that is a starting point held in common and that therefore provides a basis for attempts at resolution.[2] And as has been seen in Case 3.2, it can do this in some cases, up to and including the point of choosing an action in that case and justifying it as the best action. But how, and in what cases?

As was discussed in Chapter 3, principles shared in the common morality are general and vague; thus, the principles as they are shared in the common morality are initially unhelpful for resolving morally problematic cases. But the universal, general principles can be specified, and made more explicit and specific by that process; the resulting specified principles can be expected to be of use in addressing some difficult cases. It would be unreasonable to expect, absent any proof thereof, that a set of principles specific enough to be useful in resolving difficult cases could be shared universally; but it would not be unreasonable to expect that a specified set

---

[2] Beauchamp, Tom L. Personal communication (email), Sept 28, 2001.

of principles would be useful in resolving at least some moral conflicts. It therefore seems prudent to ask in what contexts more specified principles can be shared.

Persons will specify principles so as to make the specified principles consistent and coherent with the other beliefs that they hold. This means that when persons who hold significantly similar sets of moral beliefs specify principles, they may well be able to specify them similarly. It would be reasonable to expect that persons with similar moral beliefs would be able to utilize similarly specified principles to address and resolve difficult cases, though they will not always necessarily do so.

See, for example:

**Case 5.1: An Anencephalic Fetus** A pregnant, devout Roman Catholic, whose husband and close family are similarly devout, visits a public hospital for testing to confirm a diagnosis of a serious disorder in her fetus. The testing proceeds and it is confirmed that the fetus is anencephalic. The fetus does have a brain stem, but how much is unknown and whether it could survive, either with or without constant intensive care, is also unknown. Hospital policy suggests that an abortion may be performed in such a case. The woman asks herself whether she may morally abort this fetus. Her physician and nursing staff, who are also devout Roman Catholics, ask themselves whether they may morally perform an abortion in this situation, if asked to.

In this case, the patient, her family, and the health care team she is working with would all likely be able to specify the principles similarly. It is possible that they could all, on grounds of their similar beliefs regarding the status of fetuses, souls, and the obligations one may have towards ensouled beings, hold that it is maleficent to abort this being which deserves non-maleficent treatment. Human life, they could argue, is understood to be important, regardless of developmental level or quality of life. Because of that, the autonomous choice to abort a fetus, even a fetus that will not live long or have any quality of life, is not a valid exercise of autonomy, and based on this specification they could argue that the choice to abort cannot be morally made. They have specified principles in such a way so as to include application to all humans and to include the claim that autonomous but maleficent choices that entail the death of another human in these sorts of circumstances are impermissible.[3]

---

[3] This is not to say that all Roman Catholics will agree on the conclusions in difficult cases, even if they do share similar basic moral precepts of do good, avoid evil, don't kill, etc. But it is plausible to believe that, in this case, these persons could well similarly specify the principles, even if they are not logically required to, in order to arrive at the same conclusion in this situation. Of course, as Roman Catholics they have an obligation defined by that community to accept as authoritative the dictates of the hierarchy of the Church. As shown later in this chapter, however, that obligation has moral authority only if supported by moral concepts that the individuals involved in the case specifically accept, whether or not an organization to which they belong requires a particular resolution.

## Principle-Based Resolutions in Pluralistic Settings

When persons with significantly different rational moral views attempt to use principles to resolve morally problematic cases, it would be unreasonable to *assume* that any given set of persons would be able to succeed.[4] Nevertheless, it would not be unreasonable to suggest that, in at least some circumstances, persons with similar sets of pre-theoretical moral beliefs and even persons with rather different sets of moral beliefs would be able to use principles to help resolve difficult cases, as long as they hold sufficiently similar specifications of the relevant principles, even if the remainder of their sets of specified principles differ elsewhere.

Recall the circumstances of Case 3.2, *Refusal of Blood Products by a Pregnant Minor* as well as the discussion that followed it.[5] In that case, the resolution of the case depended importantly on the belief held by all persons involved that respecting the choice of the 15-year-old patient to refuse blood products should take precedence over providing a medically more beneficent prognosis for her. Several relevant key specifications of the principles of respect for autonomy, beneficence, and nonmaleficence were held by all of the various persons involved[6] in the case—viz., that a 15-year-old's will could be properly understood to be and respected as autonomous, that the proposed therapy was adequately beneficent even though not ideally so, that the risks of that therapy were acceptable, that the presence of a four-month fetus raised no additional concerns of nonmaleficence sufficient to mandate a different choice, and so on. These persons held significantly different views in regards to some of the questions involved in the case, such as whether it was a good idea to refuse blood transfusions, but because they were able to similarly specify principles despite those differences, they could use those specified principles to resolve the case. In the context of shared specifications, even if other moral beliefs are not shared, principles are capable of addressing and resolving difficult cases. This can happen not only when persons share significantly similar sets of moral beliefs, but also when they do not, as long as there is sufficient overlap to allow this similar use of the principles to similarly resolve the problematic case.[7] Similar arguments can apply to casuistic analyses.

---

[4] See Chapter 3, *Differing Specifications in a Pluralistic Society.*

[5] See Chapter 3, p. 93.

[6] Who is "involved" in a given case, and why those are the persons to whom the principles must be justified and by whom they must be shared, is discussed below in *Moral Acquaintanceships and the Mini-Culture of Medical Cases.*

[7] Note, too, that the amount of overlap required may be only an overlap on the specifications of the principles and not on their justifications; see *Moral Friends and Moral Acquaintances* below.

## Where Do We Go from Here?

Principles and paradigm cases are useful mechanisms for deriving solutions to moral problems where there is sufficient moral agreement underlying the moral case at hand. In cases like Case 5.1, if the persons in the case specify the principles similarly or hold similar paradigm cases to be relevant because of their similar moral beliefs, these persons can use principles or casuistry to resolve this case. It is also plausible that they may be able to resolve other difficult cases as well, on the same grounds, because their similar moral beliefs make it possible and perhaps likely that they will be able to similarly use principles and paradigm cases in other circumstances. Because these persons can be understood as being members of a moral community, one might initially think that principles and casuistry are useful only in the context of a moral community. If true, this would be a conclusion that would significantly limit the usefulness of principles and paradigm cases. If the utility of principles and cases is limited to situations where persons involved in the case share essentially similar full sets of moral beliefs, they will not be very useful; but as will be shown below, this is not the limit of the utility of principles and cases.

In Case 5.1, each person deciding about the case is a member of the same moral community, the moral community of Roman Catholicism. When persons involved in a case are all members of a given moral community, it may facilitate resolution of difficult cases; but this is not often the case in the sorts of cases discussed herein. In the modern environment of health care in most modern Western societies, the likelihood of a case involving a patient, the patient's family, health-care providers, and so on that are all members of the same moral community is low. Generally, there will be more than one moral view and/or community represented in that group of persons; sometimes there will be many. In fact, depending on how one understands moral communities, few people in a modern society may be members of a moral community at all, entirely apart from whether they are surrounded by family and physicians who are members of the same community.[8] So if principles and cases could be useful for resolving morally problematic cases only in a moral community, this would do little for resolving most problems encountered in medical ethics in a pluralistic society.

In fact, two persons' being in a moral community with each other does not entail that they hold similar sets of relevant moral beliefs, although it does usually make it more likely that they do. Within the Roman Catholic Church, there is debate at high levels about issues of theological and moral import, and at all levels from cardinals to lay members of the various dioceses, there is debate on particular moral claims. For example, despite the importance placed by the hierarchy of the church on the issue of abortion, there is dispute among members. The debate is a reason-

---

[8] See, e.g., Engelhardt's discussion of the "cosmopolitan individual" who departs from the tradition of a religious moral community on issues like abortion and contraception. If being a member of a moral community means embracing the community in full, or anything close to in full, most modern Westerners are not members of any moral community. Engelhardt, *Foundations*, 2nd ed., pp. 76–77.

able one, even considered only within the context of Roman Catholicism. In Case 5.1, for example, one could point out that Aquinas argues that the evidence of the soul in humans and not in animals is the rationality that humans possess, and which anencephalic humans do not possess.[9] Consequently, one could argue that, on this understanding of the soul, abortion in the case of an anencephalic infant is acceptable. The existence of a moral community, as they are generally understood, is no guarantee of similar beliefs regarding the issues relevant to a particular case, and thus no guarantee of an ability to similarly specify principles or choose paradigm cases. Thus, Case 5.1 could remain a morally problematic one despite the shared moral community if anyone concerned with the case differently specifies, weights, or balances the relevant principles.

Of course, if the persons involved do similarly specify the principles or choose paradigm cases, Case 5.1 can be resolved as it was suggested above, and an action taken and justified. The moral community provides no guarantee of that; but insofar as the persons involved can similarly utilize principles and paradigm cases, difficult cases can be resolved.

But what this shows is that it is not the moral community, per se, that matters in allowing persons to rationally utilize principles and cases to resolve morally problematic cases, but rather the ability of each person involved in the case to specify the principles similarly or select similar paradigm cases coherently with each individual person's corpus of moral beliefs. So, in order to discover to what extent it is reasonable to expect principles and casuistry to be able to resolve difficult cases, one needs to know to what extent the persons in a case are able to similarly specify principles and/or select paradigm cases. Since this depends upon the different persons having relevantly similar moral beliefs, if only in the areas important to resolving a particular case, it would demand both too much and too little to insist that persons be members of the same moral community in order to use principles or casuistry to resolve morally problematic cases. It is too demanding because much more is required for persons to be in moral community than simply being able to specify maxims or principles similarly in one difficult case; at the same time, it is insufficient because persons can be in moral community together and still be unable to similarly specify moral claims in a given case due to rational dispute within the community over just those issues.[10]

---

[9] See, e.g., Aquinas, St. Thomas. (1928). *Summa Contra Gentiles*, literally translated by the English Dominican Fathers. Chicago: Benziger Brothers: Third Book, Part II, Chap. CXII; see also Aquinas, St. Thomas. (1918). *Summa Theologica*, literally translated by the English Dominican Fathers. Chicago: Benziger Brothers: Part II, Question 65, Article 3.

[10] The concept of moral community has not been carefully analyzed here. This concept is complex, and any significant discussion of it would require a fair amount of nuance not included in the description above of one small group of persons who are in one form of moral community. There is a wide range of communities that have been described in the relevant literature as a form of moral community; they vary in many different ways, including the extent of moral agreement or disagreement within that community, whether there are persons in the community with moral authority to interpret moral situations, etc. The features of a moral community, including the extent to which persons in such a community can or must share moral beliefs, are almost entirely untouched here,

## Moral Friends and Moral Acquaintances

The concept of moral acquaintances, as proposed by Erich Loewy and, in a somewhat different form, by Kevin Wm. Wildes, S.J., resembles this notion of moral community in a very important way.[11] Moral acquaintanceship does not require the existence of a moral community, however that term is properly defined, yet it may be a useful device for discussing what contexts it may be reasonable to expect principles and casuistry to be able to address morally problematic cases. Because the concepts of moral friendship and moral acquaintance are initially vague, triangulation in on what the very concept of moral acquaintanceship or moral friendship means is a necessary first step.

An understanding of moral acquaintanceship is best approached through a more inclusive notion used by Engelhardt in discussing moral community: "moral friendship." This concept is meant to distinguish between persons who share some moral claims in common ("moral friends") and those who do not ("moral strangers"). Persons who are members of the same moral community, as the concept is used above, would therefore be moral friends who share a rather full conception of what morality means and requires. But the concepts are subtler than that initial phrasing might suggest. Moral friends share some moral concepts and/or some notions of what morality is and requires, though they need not share all or many such concepts.[12] They hold similar beliefs with regard to the issues with regard to which they are moral friends. So, in addition to those who are moral friends with regard to a full worldview, there are others who could also be moral friends with regard to a much narrower number of claims, yet hold other quite different moral beliefs. If two persons hold a particular moral claim to be true, then they seem to be "moral friends" on that single subject even if they are complete moral strangers on other subjects. Engelhardt implies as much when he notes that the term "moral strangers" denotes persons "who *in small or large areas* do not share a common concrete religious, moral, or philosophical viewpoint."[13] The key (italicized) phrase is elaborated when he goes on to note that "[p]eople can be both moral friends and strangers to one another...."[14] Though he holds that this happens depending on how embedded in particular moral communities these persons are, it isn't necessary to explain *how* this occurs to note

---

and they must be discussed if one is going to discuss moral community in any depth. That is not a goal of this work. The rhetorical use here of one form of moral community is intended only as a stepping stone to discussing moral friends and moral acquaintances.

[11] Wildes, Kevin Wm. (2000). *Moral Acquaintances: Methodology in Bioethics*. Notre Dame, IN: University of Notre Dame Press and Loewy, Erich. (1997). *Moral Strangers, Moral Acquaintance, and Moral Friends*. New York: State University of New York Press.

[12] See, for example, Engelhardt, H. Tristram, Jr. (1991). *Bioethics and Secular Humanism.* Philadelphia: Trinity Press International; Wildes *Moral Acquaintances: Methodology in Bioethics*; Wildes, Kevin Wm, S.J. (1997). "Engelhardt's Communitarian Ethics: The Hidden Assumptions", in *Reading Engelhardt: Essays on the Thought of H. Tristram Engelhardt, Jr.*, Brendan Minogue, Gabriel Palmer-Fernandez, and James E. Reagan, eds. Boston: Kluwer Academic Publishers, pp. 77–93.

[13] Engelhardt, *Bioethics and Secular Humanism.* p. 3. Italics added.

[14] Ibid., p. 3; *Foundations*, 2nd ed., pp. 24–25, note 13.

*that* it does. If persons are moral friends, it is because they share moral beliefs; moral strangers are strange because they do not. Thus, persons can be both moral friends and strangers if they share some, but not all, moral beliefs; and this can be true even of persons who are in very close community with a lot of agreement, such as a group of Jesuit priests and brothers. Where they are in moral agreement, they are moral friends; but where they are not, they are moral strangers. It also follows that one can be moral friends on a particular decision, such as being opposed to capital punishment, but moral strangers on the justification for that decision. All of this follows rather straightforwardly from this understanding of moral friendship.


## Some Examples

Some examples may clarify this more fully. Consider the American Civil Liberties Union (ACLU). Though Larry Flynt, publisher of various pornographic materials, and Michael Dukakis, former governor of Massachusetts and former Democratic presidential nominee, might both be members, they do not (let us suppose) share much in the way of beliefs outside of the basic beliefs of the ACLU. Yet insofar as their beliefs about the appropriate purview of freedom of speech and the press are similar, they are moral friends with regard to these views, though they may very well be moral strangers with regard to any views not related to the sorts of First Amendment questions the ACLU discusses. They may also be moral strangers on the reasons for why they believe the ACLU to be correct. One may argue that freedom of speech and press are required for a strong democracy, while the other might hold that the government of a capitalist society ought not restrict business ventures based on speech or media that some feel offensive, but should rather trust in the invisible hand of the market to quash these if they are really not desired by the society. Still, they remain moral friends on the rightness of the policies of the ACLU, even if they are moral strangers on their reasons for that rightness being justified.

Perhaps this is not so implausible. The ACLU is a group of persons who, though they may be very different elsewhere, are explicitly joined together because of a moral or at least political agreement on the importance of a strong interpretation of the First Amendment. To say that they are moral friends on the matter of the First Amendment, but perhaps not on other grounds, is surely not so surprising. Other cases of moral friendship and moral strangeness being combined in the same persons may seem more extreme, but the same features that define moral friendship in the case of the ACLU also define it in more extreme cases. For example, if the Pakistani President and the Chairman of the Joint Chiefs of Staff of the US agree on the laws of war and how they apply in the case of using chemical weapons, but do not agree on any other dimension of the ethics of war, such as with regard to precision bombing, one may say that they are moral friends on the issue of chemical weapons but not necessarily moral friends with regard to anything else. They are therefore moral friends on one issue and possibly moral strangers on all other issues. This is as much a consequence of this understanding of moral friendship as is the account of the friendship of members of the ACLU. The moral friendship in this latter case will be a fair amount less useful in addressing very many problematic cases than will the friendship in the ACLU, but that does not change the fact of their minimal moral friendship.

This suggests that moral community in and of itself is relatively morally unimportant for resolving problematic cases, though the moral agreement generally inherent in moral community is important. In effect, the notion of community becomes supplanted by the concept of moral friendship understood as moral agreement. That importance was masked above by discussion of community, though it was always the relevant important concept, and moral agreement becomes more important in understanding the relevance of moral friendship and moral strangeness.

An example from Engelhardt may further help clarify this concept of limited or minimal moral friendship. He suggests that "secularized Yuppies" may meet as moral friends on the street though they have distinct differences in their religious beliefs. The reason for their *being able* to do this may well be, as Engelhardt argues, that they compartmentalize their religious beliefs, thus failing to be as embedded in that community as others might be,[15] but the reason they *do* meet as moral friends is no more or less than that they share moral beliefs (say, for example, the rightness of a free enterprise economy.)[16] But they do not share other moral beliefs, such as perhaps those related to religious belief. Their membership in a moral community, if they are a member of one, is relevant to part of their lives, but neither important nor relevant to whether they can meet as moral friends as far as sharing basic ideas about the rightness of a free market economy. They are therefore moral friends on one issue but moral strangers on another.

They may also not believe that their shared moral beliefs are true for the same reasons, since the reasons for their shared beliefs may well extend to their unshared beliefs. One may believe a market economy is what God planned for the world; another may believe that God desires us to have a much more communal world, but one cannot morally enforce that on others; yet another might not believe in God, but merely thinks the market is the best way to make fair and just transactions. But each of these three justifiably believes that the market economy is the best sort of economy, and that it is right to maintain and act in that economy. Thus, it seems that there is a sort of moral friendship that can exist in such a case, and this friendship encompasses only the belief that a market economy is the right sort of economy, not anything else, including the reasons for why that economy is best. With regard to their religious beliefs, they are moral strangers, and for no other reason than they do not share those beliefs; with regard to their beliefs about the rightness of capitalism, they are moral friends, for no reason other than that they do share similar beliefs in this area; with regard to their beliefs about why capitalism is right, they are again moral strangers. This notion of moral friendship with regard to particular issues can

---

[15] *Foundations*, 2nd ed., pp. 76–77 and 79, referring to "ecumenical cosmopolitans."

[16] Engelhardt, *Bioethics and Secular Humanism*, pp. 3 and 35–40.

be labeled by the term "moral acquaintance."[17] Since the two persons who have independently coined it have used this term in somewhat different ways, and since it is used in a somewhat different fashion here, an analysis of the concept is in order.

•

## Three Non-rival Versions of Moral Acquaintanceships

Erich Loewy and Kevin Wm. Wildes, S.J., independently and contemporaneously developed the concept of moral acquaintanceships to fill in the gap between robust moral friends and moral strangers. Both are strongly influenced by Engelhardt's work, and though their understandings of moral acquaintances differ, their differing views can both be true at the same time.

### Erich Loewy: Moral Strangers, Moral Acquaintances, and Moral Friends

Erich Loewy argues that libertarian analyses of what a secular morality can require of moral strangers are ultimately self-defeating. As noted in Chapter 2, Engelhardt argues that the inability of religion or reason alone to define for all reasonable moral persons the content of morality means that we can only require of each other that they leave us in peace to pursue our own versions of the good life. While accepting Engelhardt's foundational claims about the inability of religion and reason to create for us all a full moral world, Loewy argues that Engelhardt still radically understates the realm of moral knowledge we can have about others, even those who are moral strangers. In fact, he argues, we know quite a bit of content about other persons, even if we know nothing about their moral views. This knowledge not only allows us to develop a fuller ethics for interactions between moral strangers but in fact requires that such a fuller ethics be developed.

Loewy's argument is as follows: we know about others at least what he calls the "existential a prioris"—six claims that are purportedly true about all persons. These are:

> (1) an inborn and necessary drive to survive ("being"); (2) the satisfaction of biological necessities (food, drink, shelter, etc.); (3) the meeting of social needs (different though they may be, all creatures have social needs); (4) a desire to avoid suffering; (5) a common sense of basic logic (at least sufficient so that the incompatibility of "p" and "non-p" are evident); and (6) a desire to lead their own lives in their own way.[18]

---

[17] See Wildes, *Moral Acquaintances*; see also Loewy, *Moral Strangers, Moral Acquaintance, and Moral Friends*.

[18] Loewy, *Moral Strangers, Moral Acquaintance, and Moral Friends*, first noted on p. 4. His language notwithstanding, it seems clear that these are not claims known a priori, but rather a posteriori. One or more of them could well not be the case for some sentient species, or even for human beings in the future. See Hanson, Stephen S. (2009). "Moral Acquaintances and Natural Facts in a Darwinian Age", in *The Normativity of the Natural*, Mark Cherry, ed. Springer Press, pp. 197–219. Nevertheless, I will continue to employ Loewy's language of "existential a prioris" here.

Loewy does not argue for the truth of these claims, because he believes that they are necessarily the case for all sentient beings. Whether or not this is true, I believe that they are clearly empirically true. One might ask for more careful analysis of the specifics of, for example, the fourth—some religions hold a value to at least certain forms of suffering—but beyond that each seems unarguably true. Because these are a part of the existence of all normal human beings, one can know that all other persons experience at least these six features.[19] They are "the 'givens' of our existence" and are "rooted in the human condition."[20] He does not see them as natural laws, but rather "demonstrable conditions"[21] which are the preconditions for ethics and indeed all human endeavors.[22] Therefore people know quite a lot of moral facts about each other; potentially, this knowledge can be interpreted into a fuller universally understood and acceptable moral theory.[23]

The six existential a prioris are also lexically ordered and interdependent.[24] Without the earlier a prioris satisfied, any purported satisfaction of later ones is shallow and insubstantial. He argues:

> To live freely and to shape our individual lives requires that we can operate with a sense of logic, that our suffering (including the suffering we have because of inadequate access to basic biological and social needs) is minimized, that our social needs (including, among others, education, health care and the freedom to pick our own friends and associates) are not stifled, our basic biological needs met, and our being underwritten. Without existence (being) the other *a prioris* are moot.[25]

The satisfaction of each higher numbered a priori is empty and pointless if the prior ones are unsatisfied. Persons regularly cease their educations in order to go to work to feed themselves and their families. Loewy argues that freedom to shape our individual lives without adequate access to food, shelter, education, etc., is not freedom at all—the freedom to freeze, starve, and die is hardly what is meant by the term "freedom."[26] Therefore, to maintain that freedom, people must ensure that the other five a prioris are fulfilled, not just for themselves, but also for others in the society as well.[27] Freedom, on his account, is necessarily social.[28]

---

[19] Indeed, Loewy points out throughout the work that not only humans experience these a prioris, but also so do most higher animals. This interesting point deserves notice, but will not be pursued here.

[20] Loewy, *Moral Strangers, Moral Acquaintance, and Moral Friends*, p. 105.

[21] Ibid., p. 87.

[22] Ibid., p. 106.

[23] As he does in his fourth chapter. See Ibid., pp. 119–172.

[24] Ibid., pp. 107 ff.

[25] Ibid., p. 117.

[26] Ibid., pp. 165–166.

[27] Ibid., pp. 171–172.

[28] Ibid., p. 171.

The existential a prioris can only be developed and maintained in a community.[29] Free and independent individuals, therefore, can only exist in the context of a community (even a hermit, he notes, exists in the context of a community[30]) just as much as communities can only exist composed of individuals. The two are conceptually connected, but just as important, they are morally connected as well.[31] In order to preserve oneself as a free and independent individual, one logically must preserve and develop the social community in which one lives.

This makes the notion of an asocial individual self-defeating. The free individual who accepts as much of society as he desires and rejects what fails to fit into his conception of the good life is a myth. For one cannot have a free individual without a broad society to support and develop that individual. Though the desires of the larger society and the desires of individuals in it may run counter to each other at times, the two are not and cannot be in fundamental opposition. Without the society, individuals will cease to be able to fulfill their needs for the satisfaction of existential a prioris, and will thus cease to be able to be free individuals.

Loewy argues further that a community that fails to provide for the basic needs of all its members soon ceases to be a viable community, and will collapse, generally sooner rather than later.[32] From this we may conclude that self-preservation requires that one must maintain the community and broader society that one exists in. Without a community, one cannot have the needs recognized in the first five a prioris, as well as the a priori of the desire for true freedom itself. Since societies do not survive without active support from their members, members of a society are obligated to support it.[33]

What this means for the understanding of moral acquaintanceship is that there are fundamental features of human existence that we all share, and from which some basic moral principles can be derived. We are not limited to permission alone but rather can build upon our basic human nature for a meaningful shared ethic, though still not one as in-depth as that of a full moral community. Loewy illustrates this with broad principles:

> ...(1) an understanding that all of us are united by a common framework of needs, interests, and capacities; (2) the necessity of keeping the peace, maintaining mutual respect, and utilizing an agreed upon process for resolving problems; and (3) the desirability of assuring full access to the basic biological and social necessities we all have and all know we have.[34]

---

[29] Here and elsewhere, Loewy uses the term community in a social sense rather than in the sense of a moral community. The community that one needs is a broad social collective that is adequate to provide the existential a prioris, which may include city, state, or even country. This might be the same as one's moral community in some rare cases—viz., the Amish societies in the Northeastern United States—but in most cases includes many who are moral strangers.

[30] Ibid., p. 23.

[31] Ibid., p. 25.

[32] Ibid., pp. 186–187.

[33] Ibid., pp. 123, 127.

[34] Ibid., p. 163.

He also elaborates (in a fashion which one might describe as specification) these more basic principles to discuss a number of social issues, arguing for particular resolutions to the issues of poverty and the welfare state, birth control, and affirmative action, among other concerns.[35]

Human beings are, on Loewy's account, all morally acquainted insofar as we all share, and know we all share, a number of "bio-psycho-social 'facts' which are unavoidable" in human existence in the world as we know it.[36] These facts lead not to an overarching, all encompassing ethic for all cultures, all world-views, and all times, but rather provide a common ground for some framework on which individual cultures can begin to develop their ethics.[37] This is a means of recognizing certain moral truths about ourselves and others that, though they allow for quite a wide variety of moral systems to develop from them, and thus much moral strangeness, also make us all morally acquainted in a narrow and basic, but undeniable, fashion.

Loewy's interpretation of the phrase "moral acquaintance" thus means one who is neither an Engelhardtian moral stranger, nor a robust moral friend, but one who shares some morally salient universally shared and universally recognized basic features of sentient/human existence with others.[38] This falls far short of a content-rich moral friendship but still is sufficient "to enable fruitful dialogue even across starkly different cultural groups."[39] Our moral acquaintanceship is necessary, based on unavoidable and undeniable facts about our existence, and thin on content without being devoid of it. That thinness does not prevent it from being useful in developing some moral claims, and even some rather specific moral and policy decisions, but it does allow the acquaintances—e.g., at least, all human persons—to develop multiple cultures and moral systems consistent with the claims of the acquaintanceship.

Though he does not employ this language, it would not be inaccurate to describe Loewy's interpretation of moral acquaintances as a defense of a kind of common morality. Since he argues that many morally serious persons, especially libertarians, fail to understand the extent to which we are all morally acquainted, this is similar to DeGrazia's common morality 2, which is the morality that all morally serious persons would agree to if they developed their considered judgments carefully.[40] This form of moral acquaintanceship is not unchanging, but it does begin with a set of stable basic claims upon which can be built shared moral understandings. It could, in other words, be used as a basis for a shared secular moral theory that would have at least some shared content.

---

[35] Ibid., pp. 188–229.

[36] Ibid., p. 87.

[37] Ibid., p. 88.

[38] Ibid., pp. 31, 105.

[39] Ibid., p. 3.

[40] DeGrazia, David. (2003). "Common Morality, Coherence, and the Principles of Biomedical Ethics." *Kennedy Institute of Ethics Journal* 13(3): 219–230; see also Chapter 3, *Two Versions of the Common Morality*.

Loewy does not implement the theory in precisely this fashion, but he does develop clear recommendations from it with analyses that could be described as comparable to specification. The existential a prioris and the broad principles that are derived from them are general, and not specific enough to necessarily immediately address problematic cases. They are a basis for beginning that discussion, and as such can serve as a much fuller shared basis for moral discussion than Engelhardtians would suggest.

Such an understanding of moral acquaintanceships has clear value for attempts to resolve moral conflicts. In such a way it can be valuable for proponents of casuistry or principle-based theories, as it could provide a potentially universally shared context for the operation of those theories, which could then become useful in determining and justifying solutions to difficult moral questions. However, the content of this acquaintanceship is still quite limited. Loewy admits and even lauds[41] the fact that the moral framework he develops is not sufficient to resolve "all, or perhaps even most, specific moral problems. . ." though he is optimistic about its ability to address a significant number.[42] Still, it would be extraordinary indeed to suggest that the existential a prioris could be extended and specified, in a universally acceptable way, such that they could resolve a complex case such as Case 3.2. Loewy's work is a significant expansion of the moral interaction of "moral strangers," but it still must be expanded upon to achieve the goal of this work. Another form of moral acquaintanceship will be necessary if the concept is to be helpful in many morally problematic cases.

### Kevin Wm. Wildes, S.J.: Moral Acquaintances: Methodology in Bioethics

Kevin Wm. Wildes, S.J., develops the notion of moral acquaintanceship in a different fashion to fill the same gap between Engelhardtian moral strangers and those who share a deep, meaningful, and complete understanding of the moral world. Wildes speaks little of the existential commonalities that we all share as humans and sentient beings. Rather, he looks to the overlap between moral communities as a means of having moral acquaintance with each other. In other words, his acquaintanceships are not necessary, but contingent. Because they are contingent upon overlap in moral views, they may be quite thin or quite robust, depending upon the amount of overlap.

Wildes understands the term "moral community" in a broad sense, encompassing a fairly wide variety of groups. He gives the example of Roman Catholics as a moral community, but also allows for the possibility of "a moral community bound together around a particular issue (e.g., the peace movement, abortion, the abolition

---

[41] He suggests that universal robust moral friendship would stifle progress and unfortunately homogenize cultures, and that perhaps it is therefore good that we are all only moral acquaintances. See Loewy, *Moral Strangers, Moral Acquaintance, and Moral Friends*, p. 105.

[42] Loewy, *Moral Strangers, Moral Acquaintance, and Moral Friends*, p. 41.

of communism or genocide, etc.)."[43] He also speaks of classic Marxists or Green Party members as candidates for moral communities, especially insofar as they share the notions of group self-identity and "excommunication" of members heretical to that identity.[44] But in other communities, like the peace movement, the social community aspect seems less important, and the moral authorities are less formal.

In Wildes's analysis of the meaning of moral community, shared moral beliefs are primary. He argues that "most moral communities can be analyzed by three factors: the content of the community's moral vision; a community's understanding of moral authority and the offices of authority; and the community's self-understanding about how it should relate to other moral communities."[45] But, he notes, there can be a "wide spectrum" of moral communities with a wide variety of understandings of these criteria, in particular the last two. Community members always share a moral vision, though the detail and depth of that vision may differ.

Wildes distinguishes between moral acquaintance at the level where persons can understand each other, but do not agree on moral claims (which he calls, "A1") and acquaintanceships where persons can not only understand each other but also share some moral agreement ("A2").[46] It is the latter which are of the most interest, since the former merely enables conversation without any conclusions derivable from that conversation. A2 acquaintances may share enough in common "to reach agreement about a particular case, principle, or policy."[47] Yet these agreements may be agreements only about the proper resolution of a case or content of a policy, without any agreement on the justification for that decision or on the theoretical/fundamental foundations for that justification. Or, moral acquaintances may share either or both of the justification and its foundations; the level of agreement possible with moral acquaintances may be complex and ambiguous.[48]

He proceeds to develop the acquaintanceship (which he also labels a "common morality") inherent in the practice of procedural ethics. He identifies four elements as the non-exclusive content of that acquaintanceship: a commitment to liberty; agreement on the rule of law; limits on the moral authority of the state and a strong separation between state, society, and community; and finally, toleration.[49] He obtains this moral content inherent in the practice of proceduralism (leading to the principle of permission) by determining what is necessary for the practice of proceduralism to succeed and assuming that anyone following that ethical decision-making method must agree with all of that content. In that context, this is a reasonable assumption. As has been shown in Chapter 2, however, it does not follow that all morally serious persons in society must accept a procedural theory, nor the

---

[43] Wildes, *Moral Acquaintances*, p. 132.

[44] Ibid.

[45] Ibid., p. 127.

[46] Ibid., p. 139.

[47] Ibid., p. 157.

[48] Ibid., p. 159.

[49] Ibid., pp. 167–173.

moral claims inherent in its exercise. Specifically, one can see where morally serious persons doubt both the rule of law and toleration, at least with regard to matters they take to be very serious. It would be a mistake—a mistake that Wildes does not clearly make, though he may—to assume that this acquaintanceship is shared by all in a liberal Western society. Wildes does argue that there is a moral acquaintanceship inherent in procedural bioethics, and that it is appropriate for bioethics in a secular, pluralist society to employ procedures to help resolve issues and identify the (minimum lower) boundaries of morality.[50] This implies that the acquaintanceship in question is one that need not be shared, though it is quite useful for persons working in bioethics to share it (and perhaps futile to work in bioethics if it is not shared.) However, he also argues that certain "practices and the[ir] corresponding moral assumptions and commitments. . .must be shared in a secular, pluralistic society," by which he means the four elements of a procedural morality noted above.[51] He also holds that, "[w]hen there are different views of the good life or what constitutes appropriate health care, consent becomes crucial for the moral justification of cooperative action,"[52] which can be interpreted to mean consent is required for moral justification, or merely that it is required for cooperative action, but one may be able to seek moral justification elsewhere for non-cooperative action.

It is also worth noting that the issues that he uses as test cases (physician-assisted suicide and abortion) can both inspire quite varied reactions.[53] Each of these issues can inspire, and has inspired, persons to passive resistance if not active, deceptive, and occasionally even violent opposition. These reactions are argued as morally justified by those engaging in them, yet are directly contrary to one or more of the four elements inherent in procedural ethics, thus placing them outside the moral acquaintanceship of procedures, at least with regard to these matters.

Wildes appears to argue that the procedures of morality are not required, but are commonly used in bioethics, and that they carry with them certain required moral principles which may be useful in developing further moral analysis. In other words, one may be, and perhaps most people are, members of the moral acquaintanceship that is entailed by proceduralism in bioethics, but one is not required to be such an acquaintance. What is interesting at this point is that Wildes has developed a very interesting concept—the notion of overlapping moral communities serving as grounds for a moral acquaintanceship—and only analyzed one possible acquaintanceship. This version of moral acquaintanceship can be made much more powerful than Wildes has so far developed it.

---

50 Ibid., p. 21.
51 Ibid., pp. 167 ff.
52 Ibid., p. 167.
53 Ibid., pp. 169–171.

## A Third Way: Specific Moral Acquaintances

Wildes's moral acquaintances share a narrow but productive band of agreement that allows many in society to appeal to something like Engelhardt's principle of permission to resolve moral conflict. There are two difficulties with accepting his interpretation as successful, however: first, as noted above, the agreement on the principles of his acquaintanceship is limited; and second, as it stands, the justification for any decisions made on grounds of that acquaintanceship is unclear. Once the latter is clarified, it will become clear that there is not one moral acquaintanceship that can be used to resolve problems, but rather that there are many. They are not shared by all persons, and some are not even shared by many; but insofar as they are shared, moral acquaintanceships can help resolve moral problems.

## Justification of the Decisions of Moral Acquaintances

To understand how this can be, note how Wildes's development of the moral acquaintanceship of procedures proceeds.[54] The moral commitments required to utilize procedures are defined without any reference to the justifications thereof, nor the foundations of those justifications. What was important was that the moral commitments were shared; persons who did share those commitments could utilize procedural ethics and agree that the conclusions derived thereby were morally justified. In practice, this is no different from the way a decision is made in, for example, a robust Orthodox Jewish community. As long as all the members of that moral community share the moral commitments of Orthodox Judaism, they can agree that the decisions derived by an appeal to those tenets of Orthodox Jewish thought are justified. They can attempt to justify many more decisions than could the proceduralists, but the method by which the decision was derived and agreed upon is similar.

The ease with which this material is discussed in Engelhardtian language may suggest that Engelhardt would agree easily with the concept as it is developed. Though there is some truth to this perception, it is perhaps misleading. While Engelhardtian theory is the progenitor in more ways than one of the concept of moral acquaintanceship, the theory, particularly as developed herein, will likely not be fully compatible with an Engelhardtian reading. Much of what moral acquaintanceship involves may resemble an Engelhardtian agreement between (partial) moral strangers; but the strongest difference comes from the source of moral justification for any decision being made. While Engelhardt would derive such justification from the necessity of peaceable agreement for moral language in a pluralistic society, the justification herein derives from the justification of the theories that underlie any agreement. Moral acquaintances justify their decisions in much the same way as do moral friends, not by appeal to an entirely separate "moral stranger morality." A clearer understanding of the way in which moral friends truly do justify any moral decisions that they share shows how this can be the case.

---

[54] Ibid., pp. 167–173.

## Similarities Between Moral Friendships and Moral Acquaintanceships

Though moral friends and moral acquaintances may initially seem quite different, in fact they are quite similar. I argue that a justification in a given moral acquaintanceship is as valid as the justifications given by moral friends; in the explication of this it will also become clear that these moral acquaintanceships can have an appeal and a utility broader than case resolution.

There are, perhaps, two competing views of what moral community and moral friendship might entail, which might be a bit bluntly described as the "liberal" and the "communitarian" views. The discussion here is not meant to be a careful examination of the liberal/communitarian debate or division, but the labels are useful and not entirely inappropriate for the task herein. The liberal interpretation understands moral friendship to be a chosen thing, which is determined by individuals who freely choose the communities to which they wish to belong. The communitarian view holds instead that our views, our moral notions, and even our choosing capacities are determined by our communities (which are social, moral, and political groupings all together) in the first place, and that the communities themselves to a large extent determine our moral friends. The key concept for an understanding of moral acquaintanceships is not to defend one version or the other, but rather to show that the "liberal" version can be compatible with the methods of justification of the "communitarian" version. For this to be true, one must separate the moral from the social function of community.

## *The Separation of the Moral and the Social*

One interpretation of moral friendship might strongly emphasize the role of the social and normative community surrounding the moral friends. Michael Sandel describes such a community as having "a common vocabulary of discourse and a background of implicit practices and understandings."[55] Community is something one "discovers," which suggests that it is less something which is chosen and more something that is historically determined and only discovered after one recognizes these influences.[56] Amitai Etzioni notes that "[c]ommunities are webs of social relations that encompass shared meanings and above all shared values."[57] Moral friends based on this sort of community would be determined in part by the social communities that they are a part of, such as Chinatown in San Francisco, or the Jewish community in a particular city or area.[58] Wildes leans in this direction in part of his description of moral community when he notes that some moral communities have

---

[55] Sandel, Michael. (1998). *Liberalism and the Limits of Justice*, 2nd ed. Cambridge University Press, pp. 172–173.

[56] Ibid., p. 150.

[57] Etzioni, Amitai. (1995). "Old Chestnuts and New Spurs", in *New Communitarian Thinking*, Etzioni, Amitai, ed. Charlottesville: University Press of Virginia, pp. 16–34. p. 24.

[58] Ibid.

the "Amish" option of withdrawing from the larger society to as much of an extent as possible. For the Amish, the social and the moral are intertwined; in the "Amish option," a withdrawal from social interaction with others outside the community is a means of preserving moral integrity. In this understanding of moral community, there is an intertwining of social and moral aspects. Even biological features might matter: some lesbian separatist groups in the 1970s, for example, rejected male-to-female transsexuals because of their being born as (and genetically composed as) a male.

Yet moral friendships and communities do not exist exclusively in such social contexts. This is suggested strongly by Wildes's reference to Ezekiel Emanuel's description of moral communities as being "shaped by a common notion of a good life."[59] What matters for membership in a community is that common notion of the good life, and not social, religious, or biological factors largely beyond one's control. (This understanding of moral friendship may still entail a strong social connection to one's moral friends, but that connection is chosen, not predefined.) That a moral friendship can exist independently of a social or political community is shown by considering the following example:

**Case 5.2 The Catholic and the Non-Catholic** Cathy follows the dictates of the Pope, seeks to guide her life by the theory of natural law, as conceived by the Roman Catholic Church, and otherwise believes and follows the moral dictates of the church. However, she has never been baptized, and there is no church near her for her to attend. Agnes, on the other hand, was baptized into the Roman Catholic Church as an infant. As an adult, she becomes agnostic, ceases to believe in the moral precepts of the church, and chooses to become a stringent believer of Nozickian moral theory. She still attends Mass on a fairly regular basis, though, and she maintains social and business ties to active members of the church.

Neither Cathy nor Agnes is a member of the community of the Roman Catholic Church, as they are both missing something critical to the understanding of the community, as seen by the church and as described above. Neither is a member of the group's "socio-normative web." But each is missing a distinctly different feature of that socio-normative connection. Cathy shares moral beliefs with Roman Catholicism, and is missing only the social connections and the strict criterion of membership into the Church (e.g., baptism). Agnes, though, still has much of the social interaction of the church, has fulfilled the requirements for membership, and has not been excommunicated. She simply shares no moral beliefs with them, except those few that may overlap between devout Catholicism and laissez-faire libertarianism. Each cannot be fully a member of the community that the church provides, but for very different reasons. The "community," then, can be divided up into two aspects, the social/political, and the moral. Cathy lacks the first, and Agnes the second.

Moreover, it is explicitly the moral aspect of the community that Cathy shares with full members of the community, and explicitly the moral aspect that Agnes fails to share. It seems clear that this ought to be relevant to determining moral community. Moral friendship entails shared moral concepts, and shared notions of

---

[59] Wildes, *Moral Acquaintances*, p. 129.

what morality is and requires. Agnes clearly could not be considered a moral friend of traditional Roman Catholics—she is much more of a moral stranger, as she shares none of their moral notions anymore—but it is much more plausible to say that Cathy is. She does share moral beliefs with them; insofar as the community is a community of shared beliefs, Cathy is a member.

Any notion of moral community and friendship being intrinsically connected with a social community would seem to preclude Cathy's being a member of the moral community that Roman Catholics have. However, a moral friendship, insofar as it is a 'moral' friendship, requires only that the friends share views about morality.[60] Views of the good life, as well as views of what is morally prohibited, allowed, or even morally relevant, fit under this description. No social connections, common practices or common goals are required—though they may be present, especially insofar as they are moral goals or activities. This commonality may be relevant to social, religious, or political community, but it is not required for moral friendship. Cathy, above, is a moral friend with Roman Catholics, even though they do not know of her existence, because she shares their moral views. Agnes is not, even though she is in their social and religious community, because she does not hold similar moral beliefs.

This may seem disconcerting or simply false, though, to those who want to hold that moral friendship should be linked with rest of the communitarian sense of community above. They would deny the claim that social aspects are not relevant for moral connection. Indeed, since part of the moral aspect of communitarian community involves special moral obligations towards other members of the community, which are not present when dealing with other persons, and furthermore since moral duties may come specifically from the social connections and desires of the community, moral friendship based solely on shared moral belief may seem to be lacking in an important way. An important part of the point of such communities is that the social and normative are tied in together, as they are in (for example) Amish life—in such a case, morality requires a certain sort of social existence. Social repercussions are an acceptable means of enforcing the moral code; and in many cases it may not be clear, nor may it matter, whether a particular rule is a social or a moral norm. (See, for example, Case 5.3 below.) Nevertheless, the social and the moral must be separated in order to be able to make clear sense of what the moral means.

## Moral Choices in a Social Context

The above objection is based mainly on the claim that moral duties can and do come from membership in a social community. Since that membership is not always freely chosen (such as membership in a family, cultural or ethnic group, or a religion that involves infant entry rituals such as baptism), moral duties are not delimited by

---

[60] The extent to which these beliefs need to be shared is an important question, which is considered below.

individual choice. Rather, some can and do come from the circumstances one finds oneself in, and those are just as much moral duties as those individually chosen. Thus, moral obligations, and thus moral friendship, cannot be separated from the broader social and political communitarian community's context.

However, the strictly moral obligations that are felt by being a member of a socio-normative community must come from the beliefs that the member being obligated has. The question seems to be this: can the social portion of the community have moral import, such that if one holds different moral views one can be seen to be morally wrong, rather than no longer a member of the community? Phrased another way, does a failure to share moral beliefs remove one from the moral community, or is it merely a moral failing on the part of the member, who is still a member of the moral community? If the latter, then moral friendship can demand more than just the moral beliefs of its members. However, as shown below, this cannot be so.

If the reasons for moral obligations do not compel a person—if those reasons are not seen as good reasons for actions—then there is no reason that the larger sense of community can give for why that person ought to obey, or even perceive, those special obligations. If the person does not believe that these obligations compel him or her to act, then the social, political, or religious aspects of the community cannot give specifically *moral* reasons for why he or she ought to. Granted, they might give social reasons—if you do not do this, you will be a social outcast—or religious reasons—if you do not behave in this way, you will not go to heaven—but these are not specifically moral reasons. Social and other reasons cannot require a particular moral response. Only moral beliefs that are compelling to the person holding them can in fact compel a person to hold a particular action or obligation to be morally correct. Consider:

**Case 5.3: Two Cultures, Two Sons** Jerry was born in the USA, the son of two recent immigrants from China. Jerry's parents maintain traditional Chinese ways, as much as possible, and remain active members in the Chinese-American business and social culture where they live. One of the elements of the traditional Chinese way of life is a strong devotion to the family, such that good adult children obey the parents in many things, including when and whom to marry. Jerry, however, chooses not to follow his parents' desires, and instead follows a different marriage path.

Tom was born in the USA, the son of two persons whose families have lived in the USA for many generations. Each of his parents has several countries of origin in their background, and they follow (to the best of their ability to discern it) American culture. One of the elements of that culture is that each person must find his or her own way in the world, and may choose it for him- or herself. Tom, however, defers to the preferences of his parents, believing that the appropriate respect for his parents requires that he obey their desires for the course of his life.

Jerry's parents can argue that he ought to obey them, but the only approbation they can bring to bear are the mores of the culture in which they live. They may (at most) reject him as a son or as a member of the society; but if Jerry rejects the claims that the culture makes on him, and accepts the social rejection that his family and society give him, there is nothing further that can be said about his duties. He was raised to believe that he had such duties, but if he now does not internalize them, and take them to be compelling for him, they do not have any necessary moral impact on him.

Tom, on the other hand, may have such moral duties, precisely because he accepts them, even though his society neither requires them nor will censure him for failing to perform them.[61] The support or lack of support for these duties that his society provides him has no impact on the actuality of those moral duties.

One may respond that it can be possible that Tom has the moral duties that he reasonably understands himself to have, since all he is doing is attributing extra responsibilities to himself. Perhaps he is simply taking as normal duties what he and his society would normally consider supererogatory duties, or even non-duties. Persons are normally allowed to give themselves extra duties, whether by choosing to enter into a job or another role with particular duties, or simply by taking duties upon themselves. These might not have been incumbent upon these persons if they had not chosen to accept them, but once accepted they can become morally obligatory. But, one could argue, how can Jerry refuse his moral duties, since the society itself can be the source of moral duties? He is not adding duties; he is denying them. How can this be justified?

The answer can be seen by looking at what recourse the group might have for making him reconsider. The Chinese-American community can argue that their community knows his actions to be morally wrong, that he is breaking with vital tradition, etc., and they can even levy on him the ultimate penalty of rejection from their community. What they cannot justifiably say is that he must, morally, choose to stay a part of the community. The social construct carries a system of duties and rewards, and the attainment of the rewards is often dependent on performance of the duties. If, however, one does not desire the rewards, including the reward of acceptance in the group itself, the duties do not carry any penalty nor any weight. The community is a system where there are penalties for failing to perform one's duties, but unlike morality, acceptance of the penalty is an acceptable path. The social group can present a member with the option of, "Do X or be rejected from the group," but allowing oneself to be rejected from the group cannot be shown by the community to be morally unacceptable. In a pluralistic society, a community determines values and duties internally, within its own boundaries; almost by definition, it cannot determine values external to its confines. It cannot, then, be necessarily immoral to choose the penalties that the community can enforce on one for failing to perform one's social duties. This shows that moral duties cannot be necessarily tied to one's social groups, as one can always, morally, accept social punishments or even leave the group. The moral restrictions take effect if one accepts them; but if one refuses them, the community cannot impose moral penalties on one who denies the moral importance of the beliefs.

---

[61] It might be thought that he can only have such moral duties to his parents if both he and his parents hold that such actions are morally required of him, since moral duties to another may only properly exist if both parties understand the duty. This may be true. However, it does not need to be examined at the moment to make the point being made here: if he and his parents accept these as moral duties, he has them; if he does not accept them, his socio-normative community cannot morally require that he accept them.

If so, then one can be moral friends with someone who is not within one's social community, since moral duties are recognized outside of social claims. If this is true, then, moral friendship ought to be understood separately from social communities; else, one will have to argue that persons who hold similar moral views but are socially separated are not moral friends, even if persons who hold different moral beliefs, but are within the same social community, are. If this is what is meant by "moral friendship," the word "moral" is meaningless in the term. If, however, "moral friendship" is understood as being solely a description of a group with shared moral interests, values, and beliefs, then moral matters are the sole central and defining feature of the concept.

This claim conflicts strongly with certain strongly held views of communitarian and Christian writers. Stanley Hauerwas argues fairly harshly against the claim that moral communities are or can be chosen by individuals.[62] He argues that the Christian moral community is identical, in some sense, to the social Christian community; and that community is one that is expressly not entered or (especially) left by choice. "To be baptized in Christ's death and resurrection is to be made part of a people, ... rendering the language of choice facile."[63] His argument is that making choice the factor that gives one entrance into the community makes the very character of Christianity unacceptably different, and thus Christians could not accept this account of moral community and, by extension, moral friendship.

However, whether or not such rites of entrance and exit are true for the religious and social aspects of Christianity, they seem to lead to thoroughly implausible results when the moral community is considered. Whether or not the baptized Christian who now holds new moral views is held to be a brother gone astray, who perhaps will be held more accountable for his actions than others who had not had the chance of baptism, his moral thinking is distinctly and perhaps fundamentally different from the community he has left. If he is still a member of a socio-normative community with the Christian community, then that community has little to do with moral friendship—or indeed morality—in any interesting sense, as its members would share little agreement on moral claims. The apostate "member" of the community might feasibly no longer share any significant moral beliefs with the other members; indeed, they may now disagree firmly on the very basics of what morality entails. Among other results of this, one can note that members of such a community would have no reason to believe they would be able to come to similar resolutions on moral questions.

---

[62] Hauerwas, Stanley. (1997). "Not all Peace is Peace: Why Christians Cannot make Peace with Engelhardt's Peace", in *Reading Engelhardt: Essays on the Thought of H. Tristram Engelhardt, Jr.*, Brendan Minogue, Gabriel Palmer-Fernandez, and James E. Reagan, eds. Boston: Kluwer Academic Publishers, pp. 31–44. Note that he is describing a very particular sort of moral community, like the communitarian one above, that necessarily involves a number of social and religious aspects. Whether or not he could accept the above conclusions about moral friendships may well depend on his response to the definition of what they may involve.

[63] Ibid., p. 35.

Whether or not Hauerwas is correct about the possibility of Christianity as a religious community in a society of radical free choice, he cannot be correct about moral communities and moral friendships being so restricted. That must be a matter of individual moral choice.

## A Point About Moral Learning and Personal History

This need not deny the claim that MacIntyre makes in *After Virtue*, that we are narrative selves who, though primary authors of our lives, exist in and cannot be understood outside of, a history and a context.[64] He echoes this idea in *Three Rival Versions of Moral Inquiry*, arguing that context can only be provided by a community, in the social sense of the term, because all learning has to take place in context. The argument is essentially that there can be no independent chooser that stands outside everything and chooses a path based on reason alone. We are raised by communities and learn from them, including learning how to be rational, in the context of the community. Thus, there can never be a rational chooser external to all communities, since even reason is only learned in context of a community.[65] Persons have pasts, and that is part of what defines them.

But this need not be denied in order to accept the conclusion above. No person can be a chooser unaffected by his past, initially indifferent to all possibilities, and educated without any bias or interpretation, but this is not necessary for a person to have the option to choose a moral path. Having a past, and a particular sort of moral and social education, does give one a particular bias in one's choices. This can be seen in the actions of those with strong cultural ties who maintain the traditions that were historically held to be socially and morally proper. One always chooses in the context of who one is, how one has been raised, and the social surroundings one has. However, a particular upbringing and the bias that provides does not make it necessary that one choose a particular way. The past can and must be interpreted by the persons whose past it is. The importance and meaning of an individual and social past can only be interpreted by the individual who, though she reasons in a context given to her by her community and upbringing, nonetheless reasons and interprets on her own.

One can react to and interpret one's history in a number of ways, each of which is an acceptable, but quite different, way of appreciating it. If two persons were brought up as American Roman Catholics, one might respond to this by embracing the faith and accepting that view of life, and its goods, that the faith gives him and working in the political system to see those goods protected. However, the other might eventually reject the faith and live his life in response to that past faith and his leaving of it. Each tells a coherent narrative of his life, and neither denies his

---

[64] MacIntyre, Alasdair. (1984). *After Virtue*, 2nd ed. Notre Dame, IN: University of Notre Dame Press, pp. 220–221.

[65] MacIntyre, Alasdair. (1991). *Three Rival Versions of Moral Inquiry.* Notre Dame, IN: University of Notre Dame Press, pp. 63 ff.

past. But each past is open to multiple different interpretations. The importance of the community and past history that one has cannot be denied, but it does not and cannot override the importance of the individual's immediate interpretation of that history. "What I am is, in a key part, what I inherit,"[66] as MacIntyre says; but what that means to a given person—what that person makes of it, and what that then makes of him or her—is, in a key part, up to one's individual interpretation.

## Moral Community, Moral Friendship, and Moral Acquaintance

If moral friendship and moral community are importantly defined by choice, how are moral claims justified in a moral community? Understanding this will show how moral claims are justified in a moral acquaintanceship as well.

Moral friendship and moral community, which can now be seen to be essentially similar, can exist in many forms. The ideal form would presumably be the widespread agreement of persons on most or all significant moral claims by use of similar reasoning from similar initial assumptions. Members of such a community could derive a very robust content-full moral theory from these shared assumptions. Of course, very few persons actually share such a complete agreement, even if 100% agreement is not required. Even many immediate families, monasteries, and enclaves do not share this type of moral agreement, and in most cases any group larger than that will be very hard pressed to reach such a goal. One could, however, think of this as the ideal for moral communities. Accepting that, how is a moral claim justified in this community? It must be justified by appeal to the shared moral beliefs in the community. These moral friends can justify moral claims to each other by agreement upon those claims and a justification that is plausible to each other.

What of a less ideal set of moral friends, who share many beliefs but do have their disagreements about a variety of issues? They would do exactly the same thing to justify a moral claim to one another: they would point to agreed upon claims and plausible justifications of those claims. Where there is a disagreement about a particular claim, they could seek to reason from some other shared points of agreement to an agreement; if that could not be found, they could not justify the action to one another. Here, though, the justification may not be the same all the way down—if there are important differences in beliefs that change the structure of the justification of a particular claim, different members of this moral friendship might justify the same claim based on a (mildly) different justification than the other.

Now consider a set of moral acquaintances, such as a group organized around a particular moral goal (typically these are political groups, but they can be based on moral beliefs.) Though they might come to the belief that the group agrees upon from a variety of different positions, they do all agree on the central issues. Even the basis for the belief in the important issue may differ, and differ importantly. Abortion rights activists may believe abortion to be a right due to the inherent sexism

---

[66] MacIntyre, *After Virtue*, p. 221.

of enforcing pregnancies on only women; they may believe it to be a right because of the inherently private nature of such a decision; they may believe it to be a right because of an understanding of the lack of the moral importance of the fetus. Particular activists might have some mix of these, but they need not share beliefs about the moral status of the fetus, sexism in society, or virtually anything else except the importance of maintaining the legal availability of abortion in their country. Nevertheless, they do share beliefs on the central important moral issue about abortion. Because of this, they share certain moral beliefs that can give them at least agreement on certain moral conclusions and a sense of moral commonality. They have a moral acquaintanceship with regard to abortion rights in their country as it currently exists. And with regard to those shared beliefs they can justify, at least to each other, certain moral claims related to the issues included in their acquaintanceship. This justification would occur in the same fashion as above—by appeal to the shared beliefs of the group.

Granted, the moral friends would be able to agree on the justifications of their decisions, while the moral acquaintances could not. But it does not follow from this point that the moral acquaintances could not derive a resolution to some questions and justify that answer as the correct answer. Consider Wildes's moral acquaintances of proceduralism. If asked why, for example, they agreed that there was no justification for the restriction of physician-assisted suicide by the state, they could point to the limits of state power, and the importance of personal liberty and toleration.[67] They might well have very different reasons for valuing each of these things, but the fact that they all value them allows them to answer a question and provide grounds for its justification, to the satisfaction of all members of that moral acquaintanceship.

If moral claims in a moral acquaintanceship are justified in this fashion, then there can be many moral acquaintanceships addressing many different moral issues. Note that both Loewy and Wildes discuss moral acquaintanceships that are shared by all persons or at least by many. But if the resolution of moral claims within a moral acquaintanceship can be justified by the involved persons appealing to their own moral system to justify an overlapping shared agreement, then there are many different moral acquaintanceships in which persons can use the same process to justify claims.

On this account, in order for two persons to be moral acquaintances, they need only share some moral claims; the acquaintanceship is defined exclusively on grounds of those shared moral beliefs. An acquaintanceship is formed by the overlap of two or more moral communities, and it exists only as far as that overlap extends. The overlap may be more or less, depending upon whose moral beliefs are overlapping. Persons are therefore not moral acquaintances "full stop," but rather are or are not moral acquaintances *with regard to a particular issue or question*. They are also properly described as morally acquainted not with all others, but rather with a given other or set of others. Roman Catholics and Southern Baptists may be

---

[67] Wildes, Moral Acquaintances, pp. 169–171.

moral acquaintances with regard to the wrongness of voluntary abortion (see Case 5.4 below), may not be moral acquaintances with regard to the moral acceptability of gambling or, perhaps, barrier contraceptive use by married couples, and may be moral acquaintances again with regard to the wrongness of assisted suicide. Even in the case of moral community members, then, one must still ask whether they are moral acquaintances with regard to the moral problem at hand. If they share moral beliefs about the matter at hand, and the acquaintanceship is broad enough to determine a shared solution, then there is adequate acquaintance to address the issue. For lack of a better term, call these "specific moral acquaintanceships."

How may a specific moral acquaintanceship function? Consider, as an example, the following:

**Case 5.4: Another Anencephalic Fetus** A pregnant (and devout) Roman Catholic, whose husband and close family are similarly devout, visits a Southern Baptist hospital for testing to confirm a preliminary diagnosis of anencephaly in her fetus. The tests confirm that diagnosis. The woman asks herself whether she may morally abort this fetus. The members of her health care team, who are Southern Baptist, ask themselves whether they can perform an abortion in this case, if asked to.

The participants in this case are not moral friends on multiple matters, including some matters of significant importance. Yet that has little to no impact on their method of addressing this case. Each participant in this case could perhaps hold that the decision to perform an abortion would be unacceptable, and each could justify their decision based on an appeal to their moral community. For how else would they justify their claim, if not by appeal to its rectitude in the eyes of their own community? The Roman Catholic would not appeal to the justificatory status nor to the coherence of the claims of the Baptist faith to justify her decision, and vice versa.

Thus, resolution and agreement can occur within the context of moral friendship or moral acquaintanceship. When persons are moral friends or acquaintances with regard to the matters regarding a particular issue, they may well be able to specify principles or employ paradigm cases in order to resolve problems; in this way resolutions of morally problematic cases will be possible if one can find a basis in moral acquaintance with regard to the relevant issues of the case between the relevant persons involved in that case.

This is, in fact, what was sought in Case 5.1 above. The members of the moral community of Roman Catholicism could presumably resolve the moral problem of whether to abort by utilizing principles, cases, or the like, in the context of their moral community. But, in fact, moral community may not serve as sufficient to resolve a moral conflict. All Roman Catholics may compose a moral community. But of course, the group that includes all Roman Catholics includes men and women, young persons and old, Americans and non-Americans, and so on. These are all differences that impact these persons' views and which may well affect their moral views such that they hold significantly different moral claims from one another in one or more areas. If one were to argue that members of the moral community of Roman Catholicism could always similarly resolve morally problematic cases because they share moral views and thus can always, for example, similarly specify and utilize principles and cases, such a claim would be rather obviously mistaken. Different Roman Catholics hold different moral views in particular situations, based in part on their various differences outside of their religion, and partly on their differing interpretations of their religion. Differences between persons' moral views, despite their membership in moral community, would make it such that they might not share the most important moral views relevant to a given question; and thus they might resolve a morally problematic case differently.

This understanding of the concept of friendship and acquaintanceship, and the use of principles and cases in the context of that acquaintanceship, is importantly different from what occurs in fuller moral communities. Members of a fuller moral community may not be able to address a particular case and resolve it unless they are also moral acquaintances with regard to the issue at hand. But if they do agree upon sufficient moral claims to address the question at hand, they can.

Therefore, rather than seeking to employ moral conflict resolution methods such as principles or casuistry only in the context of moral communities—which could mean that most people could not use principles or casuistry to resolve many moral problems, even including some who would consider themselves members of a moral community—one need look only to find moral acquaintanceships between the persons in a given case that encompass the relevant issues. In this way, principles and/or casuistry can be useful in a case like Case 5.4. In Case 5.4, it would be reasonable, assuming that all the persons involved hold the relevant parts of the moral views regarding abortion that their faiths profess, to believe that they, like the persons in Case 5.1, can similarly utilize a shared specified version of the principles to determine the action they ought to pursue in this case. That is, assuming that the persons therein are moral acquaintances, they can specify principles or appeal to cases and derive an answer to a moral problem. They can justify their own decisions based upon an appeal to their own coherent moral system, and can recognize the decisions of each of the other persons as correct based upon the coherence of the others' moral systems.

## A Challenge

There is a potential source of criticism of the justification appealed to in this third version of moral acquaintanceships. The term "moral acquaintance" describes an overlap in specific moral views when, as Wildes suggests, moral communities overlap. One might argue, though, that they ought not be allowed to appeal to the justification of claims in their own moral communities in order to justify claims that are held in common in the moral acquaintanceship. Rather, one might argue that their shared agreement must also extend to justification, and that they therefore must appeal to shared methods of justification for their moral claims held in the moral acquaintanceship. For reasons along these lines, contra the claims about justification above, the members of an Orthodox Jewish community (for example) might argue that there is a significant difference between the justification of claims in their community and by Wildes's "ecumenical" acquaintances. The Orthodox Jews share, to a large extent, agreement not only on four basic elements and the logical extensions thereof, but also on the moral and metaphysical foundations of their fully-developed moral system, with many justified and interconnected beliefs coming from those foundations. They share a robust, coherent system of beliefs, not merely four. Thus, they could justify their interpretations of moral cases in the context of a shared coherently justifiable system, which Wildes's moral acquaintances could not. The moral acquaintances in Case 5.4 could do so, though, because of the depth and breadth of their shared moral beliefs, which (let us stipulate) is sufficient to be, itself, a coherent enough system to justify moral claims. In other words, there are narrow acquaintances and then there are robust acquaintances, and there is an important difference between the two.

A first response to this line of argument is to be dubious. It is unclear why the persons involved in the decision would seek to justify their decisions based upon a shared agreement when they have perfectly good grounds for justifying their decisions in the moral claims of their separate faith communities. Each member of these overlapping communities might seek to justify the claims of that shared agreement, not by appeal to the shared agreement, but by appealing to the overall justificatory status of their individual faith community. Even if there were sufficient agreed upon claims shared by the two views to make a coherent and robust moral viewpoint, the persons involved would still hold their claims to be justified not by reference to that shared system, but by reference to the moral system they hold to be correct. Still, each could derive the same conclusion and hold it justified. This suggests that the decision can be justified if each participant in the decision can appeal to a coherent system in order to justify the decision, even if they appeal to different coherent systems. This is true even if there is satisfactory content to the overlap between their theories that the overlap itself could suffice as a coherent moral system.

Furthermore, the likelihood of decisions being justified based upon the coherence of the shared agreement fades as the shared agreement shrinks. Yet it is not clear that the grounds for accepting that claim as justified ought therefore to diminish. A specific moral acquaintanceship can be very narrow. Consider the following two cases:

**Case 5.5: Narrower Acquaintances** A pregnant (and devout) Muslim, whose husband and close family are similarly devout, visits a Southern Baptist hospital for testing to confirm a preliminary diagnosis of anencephaly in her fetus.[68] The tests confirm that diagnosis. The woman asks herself whether she may morally abort this fetus. The members of her health care team, who are Southern Baptist, ask themselves whether they can perform an abortion in this case, if asked to.

**Case 5.6: Even Narrower Acquaintances** A pregnant secular act-utilitarian, whose husband and close family think similarly, visits a Southern Baptist hospital for testing to con-

---

[68] The various schools of Islam hold varying views on the permissibility of abortion. Let us assume that this woman and her family follow a sect of Islam following the Hanafi school, which this author (who is relatively unschooled in Islamic theology) has determined holds at least that after 120 days abortion that is not to save the life of the mother is comparable to murder.

firm a preliminary diagnosis of anencephaly in her fetus. The tests confirm that diagnosis. The woman asks herself whether she may morally abort this fetus, knowing that she could carry it to term and allow at least some of its organs to be available for donation. The members of her health care team, who are Southern Baptist, ask themselves whether they can perform an abortion in this case, if asked to.

In Case 5.5, the locus of shared agreement is smaller, and may not be very coherent; in Case 5.6 it is surely far smaller and less coherent. It does not follow that the decision not to abort in each case becomes less justified. In each case, each person involved can appeal to a consistent and coherent system to justify their decisions; that the overlap between those systems decreases with regard to matters other than the decision at hand as the cases change provides no reason to think their decision is less justified. There may be a multitude of different systems of beliefs that all come to the same conclusion in a given case, but each system can successfully justify that claim to anyone who holds that system. If an action is correct in a given case, the justification for that action could well be overdetermined.

One justification may appeal more to one person, another to another; but each agrees on the right action to take and each appeals for justification to a coherent system which would suffice to call the belief justified if each were merely deciding alone. On what grounds, and at what point, does that decision become unjustified because someone else holds the same decision to be right for a different reason? The group appeals to different, and even mutually incompatible, grounds for the justification of their claims; but the successful justification of the same claim by someone else in a different fashion cannot make one's own decision less justifiable. From this one can derive the claim: "Moral acquaintances can *attain answers* to moral questions where the following claims are true: (a) each acquaintance agrees upon the same resolution; (b) each can justify that decision by reference to a coherent moral system."

This is not yet a resolution to the morally problematic question, as the emphasized text in the claim indicates. One step remains before this reaches the level of resolving moral conflicts as defined herein. All persons to whom this claim is to be justified must be able to recognize the justifications of the others as a valid justification. This can be accomplished if each person can recognize the views of each other person as being adequately consistent and coherent. If the persons involved can recognize the moral systems held by the others involved, and recognize that the decisions made in accordance with those systems are consistent and coherent with the moral system that those persons have, and can agree that the system being used by each of those others is a moral system, then the moral problem is resolved.

For example, Roman Catholics, Rawlsians, and Nozickian libertarians all could agree that the killing of an unwilling, innocent, sentient human being is generally wrong. They share little in the way of justification for these beliefs, but they do agree on the conclusion itself. They are therefore moral acquaintances with regard to the killing of an unwilling innocent normal adult human. They can also agree that a decision not to kill such a human is a justified decision if they are willing and able to recognize the decisions that each of the other persons has come to as properly justified, at least in the context of the other person's moral views. They will need to be moral acquaintances (in the sense that Wildes calls "A1"[69]) with regard to recognizing each other's moral language and a willingness to accept an appeal to a coherent moral view that is not their own as a legitimate justification.

This yields the modified claim: "Moral acquaintances can *derive justified resolutions* to moral questions where the following claims are true: (a) each acquaintance agrees upon the same resolution; (b) each can justify that decision by reference to a coherent moral system; and (c) each recognizes that each other acquaintance can justify their decision by reference to a coherent moral system."

This is a stronger claim than just an appeal to agreement. Persons might agree upon a particular solution to a problem without agreeing that it is in fact the right solution. One person might hold that, "You have a right to do this, but it is clearly the wrong thing to do." Another might think the resolution is inadequate, but the best she could hope for under the circumstances. Neither would be considered justified in that position under the theory put forth here. Nor would a person who agreed to a particular resolution because the voices in his head agreed that this was a good idea be considered justified in that decision. The grounds for the resolution must be valid for each person involved, and recognized as a plausible candidate for being valid by all persons involved.

Note also that this allows decisions to be considered justified without appeal to a shared set of specified principles, case judgments, or foundational beliefs. Different persons may derive their justifications from different coherent systems, and as long as they do derive their solution from a coherent system, such a solution is justified on this account of justification.[70] What is importantly different about Case 5.5 and especially 5.6 is that the persons therein are less likely to be using a similar set of principles, cases, or foundational beliefs. In Case 5.6, the decision not to abort is made for very different reasons by the pregnant woman and her health care team. Yet the same decision is made, and each is appealing to a coherent system to justify their decisions. As long as (c) above is also true, the fact that each has gotten to the same decision via a different path does not prevent all three of (a), (b), and (c) from applying, and therefore does not the solution to the problem from being a resolution to the problem.

Nor should this be a troubling conclusion. It would be pointless to try to justify a decision to an atheist on theistic grounds, just as it would be pointless to attempt a justification to some others without a theological appeal; thus, it is true, the same justification for a particular action will not serve to justify it to all persons, nor necessarily to all persons who agree that it is the right action in a particular case. But that action could still be justified to each of these persons, one via a theistic argument, and one via a non-theistic one. It does not follow from there being two different justifications that these persons could not recognize the same action as the

---

[69] Wildes, *Moral Acquaintances*, p. 139.

[70] This also justifies the decision based upon Beauchamp and Childress's understanding of justification, though they discuss it in the context of a common morality ostensibly shared by all. See Beauchamp and Childress, *Principles*, 5th ed., pp. 384–408.

correct resolution of a case; as long as each person holds the same action to be correct and justified, and can justify that action to all other involved persons *in some fashion* that those others can recognize as valid, that action fits the definition of a resolution.

Take an obvious example. Two persons can agree that executing an innocent person for a crime he did not commit is immoral, even if it would quell an incipient riot from persons who believe him to be guilty, and can agree to that resolution of a given case where this is in question. One might believe that this is so because the harm to the individual and to the justice system, because people will come to doubt that a system that can make such egregious errors truly gives good results, outweighs any benefit from avoiding the riot. The other might believe that it is so because punishment is only justifiable when the person in question deserves it due to prior wrongs, which this person does not. They do not share their justification for the action, yet can recognize that each holds the same action to be correct, and that each has a coherent justification that he holds to be adequate to justify the action. This should be adequate to call the question resolved, and to agree to perform the action. This question is resolved, though different justifications apply to each person.

It is also important to note that specific moral acquaintanceships do not require that all persons, or even most persons, share a particular set of moral views. One need only hold that some persons, in some cases, are specific moral acquaintances with regard to a particular morally problematic case, and share enough moral claims to allow them to address and resolve that case. Not all persons will have very many overlapping moral views, but when they do they are moral acquaintances insofar as those views overlap. It seems reasonable to believe that in a fair number of cases there could be enough overlap between the sets of moral beliefs held by the relevant persons in a case to allow them to employ their shared moral beliefs, possibly with appeal to a careful specification of principles or appeal to case judgments based on those beliefs, to be effective in addressing and resolving some morally problematic cases.

This is not as common as some would hope, but not as uncommon as it might be. Many persons are willing to understand an evaluation based upon another moral system as legitimate as long as they can understand the moral system. One difficulty is that some moral systems are often not well understood by some persons. This sort of problem can often be resolved by simple discussion to clarify the reasons why a particular decision is justified, and the basic ways in which a particular moral system in question works. This will not always work[71] but can be successful in many cases.

---

[71] For example, Peter Kreeft has argued in public lectures that utilitarianism was not a moral theory, and despite numerous arguments to show that what he meant was that it was a moral system which derived conclusions he disagreed with, he clung to his claim that it was not a moral system. It must therefore be concluded that, were he a part of a case like Case 5.6, he could not consider the result justified even if he agreed with the actions to be taken.

## Moral Acquaintanceships and the Mini-Culture of Medical Cases

A further concern with this understanding comes from the fact that not all persons are members of the specific moral acquaintanceship. As debates about such issues as abortion, physician-assisted suicide or withdrawal of nutrition and hydration show, there can be social disagreement with decisions made in particular cases. Even if patients and their families and their medical caregivers all agree upon a given resolution of a morally problematic case, others in the larger society can and, depending upon the decision being made, almost certainly will disagree with that decision. They will not, in other words, be a part of the specific moral acquaintanceship to which the patient, family and physicians will appeal in order to consider the case resolved. Yet, the argument herein is that the decision is indeed resolved and that resolution is justified. How can that be so if there is disagreement? That is, who must be a member of the specific moral acquaintanceship in order to consider a moral decision justified, and why is it that set of persons and not another?

Richard Zaner's work in clinical ethics gives an instructive guide to understand who must be a member of a moral acquaintanceship in order to understand a case as resolved. In *Ethics and the Clinical Encounter*, Zaner presents three theses in his descriptions of his method of "clinical liaison" ethics:

Thesis 1: The work of ethics requires strict focus on the specific situational definition of each involved person.
Thesis 2: Moral issues are presented solely within the contexts of their actual occurrence.
Thesis 3: The situational participants are the principle resources for the resolution of the moral issues presented.[72]

These theses, properly understood, can serve as a means to resolve the problems faced by the principle-based theorist looking to moral acquaintanceships to resolve clinical cases. For what Zaner has made clear is the extent to which clinical ethics is an intensely contextual and personal matter, requiring careful attentiveness to settings, dialogue, and interactions between the persons involved even to understand the moral matters at hand.[73] Any resolution of the case must recognize this, for only in the context of the actual case can one know the matter or resolve any moral questions that arise from the case.

The clinical encounter itself can create a "'mini-culture' composed of patient, family, health care team, and ethicist" which can delineate a fairly small group.[74] Among this group, it may well be possible (as it was in the case above) to obtain a clear understanding of the case and arrive at a sufficient agreement upon moral principles to resolve the case. Other persons external to this group might not concur, but they are not a part of this "mini-culture." What justifies the selection of this group as

---

[72] Zaner, Richard M. (1988). *Ethics and the Clinical Encounter*. Edgewater Cliffs, NJ: Prentice Hall, pp. 243–246.

[73] Ibid., p. 245.

[74] Wiggins, Osborne P. and Schwartz, Michael A. (2005). "Richard Zaner's Phenomenology of the Clinical Encounter." *Theoretical Medicine and Bioethics* 26(1): 73–87.

the appropriate cross-section of persons who must be a part of a moral acquaintance-ship in order to resolve a case is precisely their proximity to and participation in the case. These persons are properly situated to appreciate the "situational definition of each involved person" (Thesis 1), to apprehend and comprehend the moral values that are "presented solely within the contexts of their actual occurrence" (Thesis 2), and to work with the situational participants who are "the principle resources for the resolution" (Thesis 3). If "[m]oral issues are presented solely within the contexts of their actual occurrence," so, too, may be their resolutions.

If one recognizes the importance of "being in" the case itself in comprehend-ing and working with the moral issues involved in the case, it becomes clear who must be a member of the moral acquaintanceship in order to resolve a particular case. It is those who are intimately connected to the case—those in the "mini-culture" of patient, family, healthcare team, and ethicist, though in some circum-stances there could be others involved—to whom the ethical matters of the case are fully comprehensible, and from whom "the principle resources for the res-olution of the moral issues presented" must be derived. Those external to this group do not have the connectedness to the case to be able to know the "spe-cific situational definition of each involved person," and thus are importantly dis-tanced from the case. They need not be a member of the moral acquaintance-ship needed to justify a case decision by appeal to shared specified principles; the disagreement of persons outside the "mini-culture" does not render a decision unjustified.

The importance of being in the case is not always or perhaps even often recog-nized or accepted. Consider again the national fervor over the Terri Schiavo case (discussed in Chapter 1), in which persons in the government, the media, and pri-vate citizens around the country opined on the appropriate actions in determining whether her feeding tube should be removed or left in place. Many of these persons did not seem to agree that their views of the case were somehow less appropriate for guiding the treatment in this case. What reason can be given, then, for narrowing the moral acquaintanceship to the "mini-culture" noted above?

Zaner would likely argue that their understanding of the case is insufficient to consider them as part of the moral acquaintanceship, for his theses indicate an explicit awareness that the persons outside the context of the case cannot be fully aware of the moral issues that are fully presented only within that context. They are not "situational participants" in the case, and they are not "principle resources" for the resolution of the case. They may be able to discuss general moral principles relevant to the case, but cannot know how they properly apply to the particular case of Ms. Schiavo, because they have not been involved in the case. This would follow even if persons were well informed about the case; it is not the facts of the matter but the understanding of those facts that comes from involvement in the particulars of the case that makes one a part of this "mini-culture." A moral acquaintanceship developed in order to justify the decision to remove Ms. Schiavo's artificial nutrition and hydration need not include these outside protestors, no matter how strongly they might wish to impact the decision. They are not truly involved in the context of the case.

## Compatibility of the Versions

This understanding of moral acquaintances derives in part from Wildes's view, but is more inclusive of many different acquaintanceships, including very narrow acquaintanceships, and including some with a small number of members. Specific moral acquaintances have overlapping portions of their corpus of moral views, but do not have identical sets of moral views. Indeed, they may have significant differences with regard to various parts of their corpus of moral views; but insofar as they are acquaintances—that is, with regard to the matters about which they are acquaintances—they share similar moral views. Since specific moral acquaintances share a part of their corpus of moral beliefs, this shared set of beliefs can serve to ground a debate and/or a resolution to that debate on a limited set of moral questions among those acquaintances, despite the lack of full moral agreement and in the face of moral strangeness on other issues.

Specific moral acquaintances can be acquainted with regard to many different issues, as would be the Roman Catholic and Southern Baptists in Case 5.4. Or, they may be acquainted with regard to only a few matters, as might be the act-utilitarian and Southern Baptists in Case 5.6. Specific moral acquaintanceships can therefore be very broad or very tiny, encompassing many moral views or only a few. The acquaintanceship can be useful in resolving morally problematic cases on many occasions or few, or even none at all. Like Loewy's and Wildes's acquaintanceships, specific moral acquaintanceships can resolve issues only within the membership of the acquaintanceship; unlike their versions, the acquaintanceship may be very small. This leads to the likelihood of issues being resolvable for one set of persons while the same issue is irresolvable for another.

Such an interpretation of moral acquaintanceships is not incompatible with Loewy's and Wildes's versions. Loewy argues that all persons share certain basic a priori features that have clear moral import. This can be true and still allow that some persons share more agreement than just the existential a prioris, which agreement might form a moral acquaintanceship on this understanding of the term. There is no conflict whatsoever between these two, as they can function unchanged in the context of each other.

Wildes's version of moral acquaintanceship is also compatible with this third way. His moral acquaintances may be members of fuller moral communities that define their full moral lives, but recognize that the "procedures of consent" are important for moral conflict resolution in a pluralistic society.[75] If most or all persons in a secular pluralist society hold some version of proceduralism to be valid and thus are members of one broadly shared moral acquaintanceship; it is still possible— and in fact certainly so—that some subsets of the population share more agreement with each other that they may not share with the secular society's population as a whole. Persons may be Wildesian moral acquaintances with the whole of secular society and also be specific moral acquaintances with regard to a broader set of moral claims, though only with selected subsets of the population.

---

[75] Wildes, *Moral Acquaintances*, pp. 167–70 and 11–22.

Thus, these are not rival versions of moral acquaintanceship, but rather are compatible versions of a broad concept. Though there is some potential inclarity between Loewy's version, which argues that we share more than moral proceduralism, and Wildes's version that defends moral procedures as our basic tools for conflict resolution in a secular society, there is no logical conflict between the two. If indeed persons do share the six existential a prioris, there is no reason why that could not be a part of the secular moral acquaintanceship that Wildes argues exists in a secular society. Nor is there any reason why further specific moral acquaintanceships among smaller groups of persons could not also exist.

## Conclusion

The argument in this book is intended to be both critical and constructive. By beginning with Engelhardtian theory, the stage was set for the critical piece. Engelhardt's concerns about the lack of a unified, shared moral theory to which moral strangers can jointly appeal are very real, and must underlie any serious examination of resolving medical problems in a society with diversity and disagreement in its religious and moral views. Engelhardt shows that we need a moral theory to account for the interactions between moral strangers that cannot be the same as the theories that we use to resolve issues within moral communities. Moral stranger morality must, he holds, derive from content-free premises, so that it can be binding on all moral strangers regardless of the specific content of their own moral views. This sets the bar for a moral stranger morality very high, as no potentially unshared content can be allowed in the principles from which it is derived. Yet, he argues, this is the only way to avoid nihilism and to have a means to speak with moral authority amongst moral strangers.

Engelhardt derives a thin theory of moral stranger morality from two principles. The first, the principle of permission, states "Do not do to others that which they would not have done unto them, and do for them that which one has contracted to do."[76] The second, the principle of beneficence, "Do unto others what they see to be their good, as long as you also perceive it to be good and not otherwise morally wrong."[77] From these two principles springs the whole of the theory for resolution of moral difficulties between moral strangers; and though it is a rather bleakly thin theory in many ways, it is, he argues, the only theory to which one can justifiably appeal for such resolution.

But, as Chapter 2 showed, the demand for content-free principles sets the bar so high that even Engelhardt's own theory fails to meet it. His principle of beneficence allows individuals to define beneficence as they see fit, thus allowing nothing or anything at all to be understood as beneficent; in any case, it can never justify violating the permission of any other, even to be beneficent to that other. And the principle

---

[76] *Foundations*, 2nd ed., p. 123.

[77] *Foundations*, 2nd ed., p. 113; also pp. 123–124.

of permission, which forms the basis of his entire theory, is not content-free by Engelhardt's own understanding. It presupposes a contentful claim that resolving matters without recourse to violence to impose one's own view is morally appropriate. Though most persons hold this to be true in at least some cases, it is potentially unshared content; moreover, it is not only potentially unshared content. Many in a modern society explicitly and actively disagree with the permissiveness of the principle of permission, especially since taxation and utilization of funds for societal benefit is understood by Engelhardt to be a form of violence. Even if no one did actually disagree, one could; hence, the principle of permission is too demanding for Engelhardt himself.

In fact, the only principle that seems to meet his criterion of being content-free is the principle of reason-giving. This principle, which demands of anyone making a moral claim that they be able to give some reason for that claim, contains only the content (if that is even the appropriate term for it) that one engage in ethical discussion of justifications for action. Since the system is meant to give persons trying to speak with moral authority in the context of dealing with moral strangers a means by which to justify moral actions, this "content" is presupposed in the very action being discussed. It is, then, free of any contestable moral content for any persons to whom Engelhardt intends to speak.

This success at attaining a content-free principle of moral stranger morality comes at a dreadful cost. The principle of reason-giving is a dismally poor principle for deriving moral conclusions. It can defend very few positive moral claims, and can exclude as unacceptable very few moral claims as well. It could not even suffice to criticize the actions of a pirate captain apportioning slaves according to the rules of the Pirate's Creed, or the "consistent Nazi" who can provide clear logical arguments for his positions. If the best a moral stranger morality built on content-free principles can do still allows for the possibility of the self-contradictory term "Nazi ethics," then perhaps it is time to let go of the hope of an ethics based on content-free principles.

But even if one does this, one must still have some means for dealing with the resolution of moral problems between moral strangers. The inability to address them via content-free principles will not make the conflicts go away. What, then, is one to do? One might follow the lead of other, non-Engelhardtian, bioethicists who still take pluralism in ethics seriously and try to find a source of agreed-upon content that even moral strangers may appeal to for resolution of moral problems. Principle-based theories, like that proposed by Beauchamp and Childress, look to principles shared in the common morality for this agreement; casuists like Jonsen and Toulmin seek it in decisions about paradigmatic cases. However, since Chapters 3 and 4 show that this agreement, understood in a society-wide sense, is largely trivial or illusory, neither of these appeals will suffice as a means of resolution of moral problems. Any agreement on paradigmatic cases or specifiable principles is too vague at the society-wide level to be of any use in resolving morally difficult cases.

So Chapter 5 begins with the reader again at the brink of surrender to nihilism. Content-free principles leave us with desperately little with which to address the pressing problems of health care ethics, and the main sources of medical ethical decision-making content do not provide universally acceptable, useful, content. But the conclusions about principle-based theories and casuistry were not that they were incapable of providing tools for resolution of moral problems, but that they were limited in what problems they could resolve and to whom those resolutions could apply. These methods can work well within the context of some mutual agreement, even if that agreement is relatively basic. Principles and/or casuistic analyses can build upon even a relatively small amount of moral agreement to develop into systems that can overlap at the relevant case in which a decision must be made, even if those systems do not overlap elsewhere. Moral acquaintances can be able to use principles and cases to satisfactorily and justifiably give a moral answer to, and thus resolve, morally problematic cases, if their acquaintanceship allows them enough agreement in order to create an overlap of their moral systems sufficient to justify a decision in the case at hand. It follows from this that not all groups of persons will be able to resolve all cases; some groups could resolve some, while other groups could not resolve the same case. Similarly, a change in the persons involved in a case may allow for a decision to be justified when a decision could not be justified to the original set of persons. Moral justification is justification to a given group of persons, and not universal justification.

This is at once both a worrying and a liberating conclusion. It is worrying in that it removes the option of knowing that moral problems can be resolved by appeal to universal principles, content-free theories, or widely recognized objective truths. These claims simply cannot be justified in a morally pluralistic culture, and that result may well make the sort of moral agreement possible between moral acquaintances suspect or even thought to be non-moral by who want to hold that moral decisions must have the weight of universally recognized truths behind them. Yet, in a secular pluralistic society, no more than this can be justified.

Yet this is also a liberating conclusion, because it allows even persons who take very seriously the Engelhardtian challenge of pluralism to employ many of the methods of modern decision-making in medical ethics. The tools of principles and paradigm cases need not be discarded; they must merely be used with a closer eye to the particulars of the individuals involved in a particular case. One cannot obtain universal moral resolutions, or even universal agreement on the content of four basic principles; but one can address particular cases in the context in which they occur. Since this is what is desired by most persons seeking to resolve morally problematic cases, the conclusion that this is possible, in at least some and perhaps many cases, is extremely important.

How, then, shall an ethics consultant, or an ethics committee, or an ethical physician seek to use this conclusion to help when problems arise? The liberating conclusion allows these persons to work with many of the same tools that have been used before, but care is required. These tools will be subtly (or perhaps not so subtly) different in different cases, and discussions about prior cases and prior resolutions will have to be carefully worked through with the participants in a given case, as the same resolutions and tools may not apply in the same way in the new case, even if the medical facts of the case are entirely identical.

Consider, as an example, the following case:

**Case 5.7: No Longer Acquaintances** A pregnant secular humanist, whose partner and close family think similarly, visits a Southern Baptist hospital for testing to confirm a preliminary diagnosis of anencephaly in her fetus. The tests confirm that diagnosis. The woman asks herself whether she may morally abort this fetus. The members of her health care team, who are Southern Baptist, ask themselves whether they can perform an abortion in this case, if asked to.

In this case, where the relevant medical facts are the same as discussed in multiple cases above, the participants now hold sets of moral views that may render the case irresolvable. The pregnant woman in this case may consider certain principles of secular humanism as central to her understanding of the case, such as commitments to reason as means of understanding the world, and the valuing of human intelligence, to understand that her fetus is lacking a crucial part of its essence as a human, such that it will not be able to obtain any of the features of human life that make it valuable and human.[78] Though she may also recognize the importance of tolerance of other ways of life, and even of actively supporting persons with various mental or physical challenges so that they can be able to help themselves, she may plausibly conclude that this fetus is forever lacking any ability to have a way of life or to help itself in the relevant fashion. Consequently, she may choose at this point to abort the pregnancy, as her fetus can never develop as she would hope it could and has not got the moral standing to require that she continue the pregnancy if she chooses not to.

Her Southern Baptist health care team may not be able to join her in that moral analysis. They may recognize the fetus as having the moral status of all human beings, and believe it morally inappropriate to perform an abortion, instead believing that completion of the pregnancy to delivery is morally appropriate. Aggressive treatment of the infant after delivery need not be provided, but abortion is not acceptable here. This case may not be amenable to resolution by these persons, as their disagreements may be fundamental enough to prevent them from arriving at an overlapping agreement. In the language of principles, the health care team understands the fetus as a being to whom strong duties of nonmaleficence apply, which are not overridden by its physical abnormality, while the pregnant woman may hold that this fetus is not an appropriate target of such nonmaleficence, in particular when a significant violation of a person's autonomy is required for provision of that nonmaleficence. In the language of casuistry, the health care team may be viewing the case as a paradigmatic case of killing a human being, with the anencephaly insufficient reason to consider the case relevantly differently; the pregnant woman may view it as a case of ending a life that could never be fully human, or perhaps as providing a requested medical treatment, with the fact that the requested treatment results in the death of a brainless human fetus insufficient reason to consider the case relevantly differently. The tools of casuistry and principle-based decision-making may not allow these persons to arrive at a justified resolution to this case; a moral acquaintanceship on these grounds may not exist.

---

[78] See, for example, "The Affirmations of Humanism: A Statement of Principles." http://www.secularhumanism.org/index.php?section=main&page=affirmations (accessed April 16, 2008).

However, the methods discussed herein do not leave us without tools with which to try to resolve the case. Realizing that resolutions and justifications are resolutions and justifications for a particular group, a possible means for resolving this case is changing the membership of the group. If a physician cannot participate in this procedure because she understands it to be morally unjustified, she can still assist in resolving the issue by means of transferring the patient to a physician who holds other moral views that may allow for a morally justified solution to the problem.[79] The different "mini-culture" obtained by differing health care providers may allow for a moral resolution to the case; in the same way that the same facts with different relevant participants to the case made it irresolvable in the original version of Case 5.7, another change of participants may make it resolvable.

One may object that this places an obligation on the original physician in Case 5.7 that may be unacceptable according to her moral views. On grounds of a strong reading of a right to conscientious objection, a physician might argue that she is not only not obligated to perform an abortion that she feels to be immoral, but is also not obligated to refer a patient for such an abortion.[80] In strictly secular pluralistic terms, the physician has the right to do this. There is not an argument that can be justified in a secular pluralistic context that can require that the physician do this; however, what the physician is doing when refusing to refer the patient is also refusing to continue an attempt to resolve the case in a (secular, pluralistic) morally justifiable fashion. Moral acquaintanceships provide persons with means by which to arrive at such justifiable resolutions if they desire, as many do, but cannot require one to seek to do so. By making such an objection, the physician is here leaving the realm of those who seek justifiable resolutions of moral claims to those who are not moral friends, and so is leaving behind the goal of this work.

If, however, one desires to obtain such morally justifiable resolution, the tools to do so are available. Referral of the patient to providers more likely to be able to share a relevant moral acquaintanceship with the patient allows for this resolution; a refusal to do so does not.

The reader may note that this is largely what is done already in such cases, and may question what, if anything, the account of moral acquaintanceship has added. This account adds justification—that is, it provides reasons why one ought to act in this way, and shows why such behavior is not only consistent with common practice, or the recommendations of the International Federation of Gynecology and Obstetrics (FIGO),[81] but is also morally appropriate. This practice allows for the

---

[79] It would also be acceptable to discuss the matter between persons who disagree to discover whether an area of agreement can be found or whether one or more members of the group may change their minds. But if that were to happen, the case would then more resemble one of the above cases; therefore, in this case it is taken as a given that these methods have either been rejected or have been tried without success.

[80] For example, Pellegrino has argued for this explicitly, claiming that requiring the physician to refer is requiring her to participate in a morally objectionable act. See Pellegrino, Edmund D. (2002). "The Physician's Conscience, Conscience Clauses, and Religious Belief: A Catholic Perspective." *Fordham Urban Law Journal* 30(1): 221–244.

[81] See "FIGO Professional and Ethical Responsibilities Concerning Sexual and Reproductive Rights", at http://www.figo.org/Codeofethics.asp (last accessed April 16, 2008,) which states in

possibility of moral resolution to difficult moral cases in a pluralistic society; refusing to do so does not. Moreover, it is also clear why the decisions made by the patients and the physicians to whom they are referred are morally justifiable, as they are justified by and to the members of the moral acquaintanceship formed in the "mini-culture" of that interaction between patient, health care team, and any other relevant persons (family members, etc.) Finally, it provides an argument for why the refusal to refer patients to providers willing to perform the abortion is problematic, in that this refusal takes one outside the realm of persons seeking resolution of morally difficult cases in a pluralistic society. The claim that resolution of such cases is important could ground, for example, the FIGO policy requiring referral.

Other cases can help elaborate some of the values of approaching cases from the standpoint of seeking moral acquaintanceships. Consider the following case, which is fictional but likely recognizable to many clinicians or clinical ethicists:

> **Case 5.8: Cousin Bob** Ms. J is a 74-year-old woman in a declining state of multiple system organ failure. She has no living will or durable power of attorney for health care, and has been incapable of communication since her admission to the hospital three weeks ago. However, she has several family members who visit regularly, seem very concerned for her, and who have been making decisions for her medical care. She had been placed on a ventilator ten days ago when her breathing rate had dropped too low to keep her alive without mechanical assistance. At that time, there was some hope that, with ventilatory assistance, she would be able to take a turn for the better, regain consciousness, and leave the hospital; unfortunately, she instead took a turn very much for the worse. Further treatment of any sort now seems very unlikely to allow her to either regain consciousness or to leave the hospital alive. After multiple conversations with the physicians and other providers caring for her, Ms. J's family has come to the conclusion that, as far as they can ascertain, it would not be Ms. J's wish to continue aggressive treatment with such a poor prognosis. They discuss removal of the ventilator, which is expected to lead to a rapid decline and then death, and agree that the ventilator treatment should be stopped. They plan to do so when the resident they have been dealing with for most of Ms. J's stay returns in the morning. Before the morning can come, however, Cousin Bob arrives from Aruba with a strong message. He desires everything to be done for Aunt J, and is quite vocal and forceful in his insistence that continuing her life and aggressive therapy. When the resident arrives in the morning, Bob is the first person to encounter him, and demands that the ventilator stay in place for as long as Ms. J can survive.[82]

Where once there was an agreement upon the right action to take in this case, now there is not. Prior to Bob's arrival, the patient's family, health care providers, and (to the best of anyone's understanding) the patient herself were moral acquaintances insofar as the decision to discontinue ventilation was concerned. How is the

---

part that members of FIGO member societies will "*Assure* that a physician's right to preserve his/her own moral or religious values does not result in the imposition of those personal values on women. Under such circumstances, they should be referred to another suitable health care provider." (emphasis in original) It also goes further to note that, "Conscientious objection to procedures does not absolve physicians from taking immediate steps in an emergency to ensure that the necessary treatment is given without delay."

[82] "Cousin Bob"-type cases are discussed in Doukas, David J. and William Reichel. (2007). *Planning for Uncertainty: A Guide to Living Wills and Other Advance Directives for Health Care*, 2nd ed. Baltimore: John Hopkins University Press, pp. 67, 89, 94.

resident to approach this case? Cousin Bob is not a moral acquaintance with the rest of the family and health care team; can the decision to discontinue treatment be justified? Conversely, though, can the decision to continue treatment be justified, either? There is disagreement on both options, and no acquaintanceship can be found to encompass all persons.

Yet, again, in the concept of moral acquaintances are the tools needed to best address this case. Cousin Bob is not a member of the moral acquaintanceship that agrees that it is appropriate to discontinue Ms. J's ventilator, but it is possible that he could become so. If he were to become better informed about the case, and to learn more about her prognosis and history in this hospital stay and her prior statements that indicated her preferences, he might determine that it would be more appropriate to remove the ventilator. Conversely, were Bob to provide clear documented evidence that Ms. J had explicitly requested such treatment due to a belief, unknown to the other family members, in the inherent spiritual importance of all human life, they might all come to agree that respect for these autonomously derived beliefs should require them to continue treatment despite the poor prognosis. So our first task should be to see if the moral acquaintanceship could include Bob.

However, if Bob is not amenable to discussion, and provides no additional input into the decision outside of a desire to continue aggressive treatment, the family and health care team may proceed with the removal of the ventilator on the same grounds that they did before. Bob is not a moral acquaintance with them on this matter, but he is also not one of the "situational participants" in the decision. He has not been able to apprehend the facts of the case in the context of their occurrence, and is not a part of the "mini-culture" that this case had created among the participants who had been there the whole time. His inability to share the moral acquaintanceship that justifies removal of Ms. J's ventilator does not prevent that decision from being morally justified. He is not one of the participants to whom the decision needs to be justified. He may be vociferous, he may raise a legal case, but he cannot make the decision morally unjustified.

Other changes in practice can follow from the understanding of moral acquaintanceships that helped inform this case. The pursuit of legally authorized next-of-kin for decision-making purposes should be more limited than it often is in difficult cases. If a patient is incompetent to make her own decisions, and has not legally authorized a surrogate, we ought not seek to contact a half-sister who lives in a distant country and last spoke with the patient 30 years ago. Her input, even if it has some legal standing, is irrelevant to morally determining what to do. She is not a part of the "mini-culture" participating in the case, and would be ill-prepared to become part of it.

Similarly, when persons are situational participants, their participation in the decisions involved is required in order for the decisions to be morally justified. This explains why it was so inappropriate for long-term partners of dying gay men in the early years of the AIDS crisis to be denied the ability to assist in decisions with regard to their loved one, while families who had been estranged from their son since he'd come out to them 20 years before were given the ability to determine treatment. This is a complete reversal of the appropriate arrangement of the case. If an acquaintanceship between all persons cannot be arrived at, it is the partner who has been a part of the case as the disease developed, and not the family, who must be a member of the moral acquaintanceship that addresses the decisions in the case.

There are also implications of moral acquaintanceships for hospital policies. Consider, for example, a policy regarding futile (or non-beneficial) treatments. Such a policy would have to require that there be adequate acquaintance amongst the relevant participants in a case before one can determine that a treatment is non-beneficial. For what benefit is, is determined at least partly by the persons asking the question. Whether continued comatose existence, or lessened length of survival in exchange for greater lucidity, or a small chance at partial remission of a life-threatening tumor, and so on, is truly futile or non-beneficial is determined largely by the values of the persons involved in that decision.[83] The patient and his or her close participants in the case need to agree that the proposed treatment provides nothing of value to the patient before a treatment can be discontinued as "non-beneficial" or "futile." Physicians cannot make such a claim unilaterally.

The rationale behind this claim is that declarations of futility are in part moral claims. There is always a significant factual portion of the claim, of course, and sometimes that is a large part of it. Claims that, for example, certain alternative therapies have not been shown to have any noticeable effect are certainly factual claims; however, the further claim that, "This alternative therapy in this case is not worth providing," is a value statement. Consequently, decisions about whether a particular treatment is beneficial will very often be moral decisions. Such decisions must be made in an acquaintanceship between the patient and health care providers, as well as all other relevant participants to a decision.

These few comments are not meant to circumscribe the limits of the implications of moral acquaintanceships, nor are they any more than a beginning indication of what should follow. Hopefully, however, they can serve as a means to assist in exploring the implications of doing health care ethics in the context of a recognition of the importance of Engelhardt's challenge. We are not damned to nihilism by the challenges of secular pluralism, nor are we constrained by the limitations of pure content-free "near-nihilism." In fact, by employing them within the context of moral acquaintanceships, we can make use of many of the tools that modern bioethical practice has already developed. Though we must make a more limited, more careful, and more subjectively determined use of these tools, that does not prevent us from resolving morally problematic cases, addressing moral disagreements, or even making policies with moral justification. Moral acquaintanceships open the discussion of "moral stranger morality" to "moral acquaintance morality," and, in so doing, allow a much fuller practice of health care ethics in a modern society.

---

[83] It is possible that some of the most egregious cases of true futility may be purely scientific in nature—e.g., casting an unbroken bone provides no physiological benefit. But in morally problematic cases of futility, which includes most of those discussed in any depth, a value judgment needs to be made.

# Bibliography

Aquinas, St. Thomas. (1918). *Summa Theologica*, literally translated by the English Dominican Fathers. Chicago: Benziger Brothers.

Aquinas, St. Thomas. (1928). *Summa Contra Gentiles*, literally translated by the English Dominican Fathers. Chicago: Benziger Brothers.

Arras, John. (1991). "Getting Down to Cases: The Revival of Casuistry in Bioethics." *Journal of Medicine and Philosophy* 16: 29–51.

Aulisio, Mark P. (1998). "The Foundations of Bioethics: Contingency and Relevance." *Journal of Medicine and Philosophy* 23: 428–438.

Baier, Annette C. (1994). "Trust and Antitrust", in *Moral Prejudices: Essays on Ethics*. Cambridge, Mass: Harvard University Press.

Beauchamp, Tom L. (1987). "Ethical Theory and the Problem of Closure", in *Scientific Controversies: Case Studies in the Resolution and Closure of Disputes in Science and Technology*, H. Tristram Engelhardt, Jr. and Arthur L. Caplan, eds. New York: Cambridge University Press, pp. 27–48.

Beauchamp, Tom L. (1994). "The Four-Principles Approach", in *Principles of Health Care Ethics*, Raanan Gillon, ed. New York: John Wiley & Sons, pp. 3–12.

Beauchamp, Tom L. (2000). "Reply to Strong on Principlism and Casuistry." *Journal of Medicine and Philosophy* 25(3): 342–347.

Beauchamp, Tom L. (2001). *Philosophical Ethics*, 3rd ed. Boston, MA: McGraw-Hill.

Beauchamp, Tom L. (2003a). "Methods and Principles in Biomedical Ethics." *Journal of Medical Ethics* 29(5): 269–274.

Beauchamp, Tom L. (2003b). "A Defense of the Common Morality." *Kennedy Institute of Ethics Journal* 13(3): 259–274.

Beauchamp, Tom L. and James F. Childress. (1994). *The Principles of Biomedical Ethics*, 4th ed. New York: Oxford University Press.

Beauchamp, Tom L. and James F. Childress. (2001). *The Principles of Biomedical Ethics*, 5th ed. New York: Oxford University Press.

Brody, Baruch. (1992). "Special Ethics Issues in the Management of PVS Patients." *Law, Medicine & Health Care* 20: 104–115.

California Supreme Court. *Tarasoff v. Regents of the University of California*. 131 California Reporter 14. Decided July 1, 1976.

Callahan, Dan. (1996). "The Goals of Medicine: Setting New Priorities." *Hastings Center Report* 25: S1–S26.

Clouser, K. Danner. (1995). "Common Morality as an Alternative to Principlism." *Kennedy Institute of Ethics Journal* 5(3): 219–236.

Clouser, K. Danner and Bernard Gert. (1990). "A Critique of Principlism." *Journal of Medicine and Philosophy* 15: 219–236.

Clouser, K. Danner and Bernard Gert. (1994). "Morality vs. Principlism", in *Principles of Health Care Ethics*, Raanan Gillon, ed. New York: John Wiley and Sons, pp. 251–266.

S.S. Hanson, *Moral Acquaintances and Moral Decisions*, Philosophy and Medicine 103,      167
DOI 10.1007/978-90-481-2508-1_BM2, © Springer Science+Business Media B.V. 2009

Crigger, Bette-Jane, ed. (1998). *Cases in Bioethics: Selections from the Hastings Center Report*, 3rd ed. New York: St. Martin's Press.

DeGrazia, David. (1992). "Moving Forward in Bioethical Theory: Theories, Cases, and Specified Principlism." *Journal of Medicine and Philosophy* 17: 511–539.

DeGrazia, David. (1996). *Taking Animals Seriously*. New York: Cambridge University Press.

DeGrazia, David. (2003). "Common Morality, Coherence, and the Principles of Biomedical Ethics." *Kennedy Institute of Ethics Journal* 13(3): 219–230.

Doukas, David John and William Reichel. (2007). *Planning for Uncertainty: A Guide to Living Wills and Other Advance Directives for Health Care*, 2nd ed. Baltimore: John Hopkins University Press.

Dresser, Rebecca. (2005). "Schiavo's Legacy: The Need for an Objective Standard." *The Hastings Center Report* 35(3): 20–22.

Engelhardt, H. Tristram, Jr. (1986). *The Foundations of Bioethics*, 1st ed. New York: Oxford University Press.

Engelhardt, H. Tristram, Jr. (1988). "Foundations, Persons, and the Battle for the Millenium." *Journal of Medicine and Philosophy* 13: 387–391.

Engelhardt, H. Tristram, Jr. (1991). *Bioethics and Secular Humanism*. Philadelphia: Trinity Press International.

Engelhardt, H. Tristram, Jr. (1996). *The Foundations of Bioethics*, 2nd ed. New York: Oxford University Press.

Engelhardt, H. Tristram, Jr. (1997). "The Foundations of Bioethics and Secular Humanism: Why is there No Canonical Moral Content?", in *Reading Engelhardt: Essays on the Thought of H. Tristram Engelhardt, Jr.*, Brendan P. Minogue, Gabriel Palmer-Fernandez, and James E. Reagan, eds. Boston, MA: Kluwer Academic Publishers, pp. 259–285.

Engelhardt, H. Tristram, Jr. and Kevin William Wildes. (1994). "The Four Principles of Health Care Ethics and Post-Modernity: Why a Libertarian Interpretation is Unavoidable", in *Principles of Health Care Ethics*, Raanan Gillon, ed. New York: John Wiley and Sons, pp. 135–147.

English, Jane. (1979). "What Do Grown Children Owe their Parents?", in *Having Children: Philosophical and legal Reflections on Parenthood*, Onora O'Neill and William Ruddick, eds. New York: Oxford University Press.

Etzioni, Amitai, ed. (1995). *New Communitarian Thinking*. Charlottesville: University Press of Virginia.

Feigenbaum, Frank, Sulmasy, Daniel P., Pellegrino, Edmund D., and Henderson, Fraser C. (1997, September). "Spondyloptotic Fracture of the Cervical Spine in a Pregnant, Anemic Jehovah's Witness: Technical and Ethical Considerations." *Journal of Neurosurgery* 87(3): 458–463.

Gert, Bernard, Charles Culver, and K. Danner Clouser. (1997). *Bioethics: A Return to Fundamentals*. New York: Oxford University Press.

Gert, Bernard, Charles Culver, and K. Danner Clouser. (2000). "Common Morality versus Specified Principlism: Reply to Richardson." *Journal of Medicine and Philosophy* 25(3): 308–322.

Green, Ronald. (1990). "Method in Bioethics: A Troubled Assessment." *Journal of Medicine and Philosophy* 15: 179–197.

Grisez, Germain, Joseph Boyle, John Finnis, and William E. May. (1988). "Every Marital Act Ought to be Open to New Life: Toward a Clearer Understanding." *Thomist: A Speculative Quarterly Review* 52: 365–426.

Hanson, Stephen S. (2007). "Libertarianism and Health Care Policy: It's Not What You Think It Is." *Journal of Law, Medicine, and Ethics* 35(3): 486–489.

Hanson, Stephen S. (2009). "Moral Acquaintances and Natural Facts in a Darwinian Age", in *The Normativity of the Natural*, Mark Cherry, ed. Philosophical Studies in Contemporary Culture 17, Springer Press.

Hauerwas, Stanley. (1997). "Not All Peace is Peace: Why Christians Cannot make Peace with Engelhardt's Peace", in *Reading Engelhardt: Essays on the Thought of H. Tristram Engelhardt, Jr.*, Brendan Minogue, Gabriel Palmer-Fernandez, and James E. Reagan, eds. Boston: Kluwer Academic Publishers, pp. 31–44.

Hughes, Duncan B., Brant W. Ullery, and Philip S. Barie. (2008). "The Contemporary Approach to the Care of Jehovah's Witnesses." *Journal of Trauma-Injury Infection & Critical Care* 65(1):237–247.

Jonsen, Albert R. (1991, September–October). "Of Balloons and Bicycles: The Relationship between Ethical Theory and Practical Judgment." *Hastings Center Report*, 14–16.

Jonsen, Albert R. (1991). "Casuistry as Methodology in Clinical Ethics." *Theoretical Medicine* 12: 295–307.

Jonsen, Albert R. (1992). "Casuistry: An Alternative or Complement to the Principles." in Albert R. Jonsen, Mark Siegler, and William Winslade. *Clinical Ethics*, 3rd ed. New York: McGraw Hill.

Jonsen, Albert R. (2000). "Strong on Specification." *Journal of Medicine & Philosophy* 25(3): 348–360.

Jonsen, Albert R. and Stephen Toulmin. (1988). *The Abuse of Casuistry*. Berkeley: University of California Press.

Katz, Jay. (1984). *The Silent World of Doctor and Patient*. New York: The Free Press.

Kerridge, Ian, Michael Lowe, Michael Seldon, Arno Enno, and Sandra Deveridge. (1997). "Clinical and Ethical Issues in the Treatment of a Jehovah's Witness with Acute Myeloblastic Leukemia." *Archives of Internal Medicine* 157(15):1753–1757.

Koch, Tom. (2005). "The Challenge of Terri Schiavo: Lessons for Bioethics." *Journal of Medical Ethics* 31(7): 376–378.

Kuczewski, Mark G. (1994). "Casuistry and its Communitarian Critics." *Kennedy Institute of Ethics Journal* 4(2): 99–116.

Kuczewski, Mark. (1998). "Casuistry and Principlism: The Convergence of Method in Biomedical Ethics." *Theoretical Medicine and Bioethics* 19: 509–524.

Levine, Melvin D., Lee Scott, and William J. Curran. (1977). "Ethics Rounds in a Children's Medical Center: Evaluation of a Hospital-Based Program for Continuing Education in Medical Ethics." *Pediatrics* 60: 205.

Little, Margaret O. (Forthcoming). *Intimate Duties: Re-Thinking Abortion, the Law, & Morality*. New York: Oxford University Press.

Loewy, Erich H. (1987). "Not By Reason Alone: A Review of H. Tristram Engelhardt, Foundations of Bioethics." *Journal of Medical Humanities and Bioethics* 8(1): 67–72.

Loewy, Erich. (1997). *Moral Strangers, Moral Acquaintance, and Moral Friends*. New York: State University of New York Press.

MacIntyre, Alasdair. (1984). *After Virtue*, 2nd ed. Notre Dame, IN: University of Notre Dame Press.

MacIntyre, Alasdair. (1991). *Three Rival Versions of Moral Inquiry*. Notre Dame, IN: University of Notre Dame Press.

Miller, Franklin G. (1998). "The Internal Morality of Medicine: Explication and Application to Managed Care." *Journal of Medicine and Philosophy* 23(4): 384–410.

Miller, Franklin G. and Howard Brody. (2001). "The Internal Morality of Medicine: An Evolutionary Perspective." *Journal of Medicine and Philosophy* 26(6): 581–599.

Minogue, Brendan P., Gabriel Palmer-Fernandez, and James E. Reagan, eds. (1997). *Reading Engelhardt: Essays on the Thought of H. Tristram Engelhardt, Jr*. Boston, MA: Kluwer Academic Publishers.

Moskop, John C. (1997). "Persons, Property, or Both? Engelhardt on the Moral Status of Young Children", in *Reading Engelhardt: Essays on the Thought of H. Tristram Engelhardt, Jr.*, Brendan P. Minogue, Gabriel Palmer-Fernandez, and James E. Reagan, eds. Boston, MA: Kluwer Academic Publishers, pp. 163–174.

Murray, Shailagh and Mike Allen. (2005, March 26). "Schiavo Case Tests Priorities of GOP." *The Washington Post*, A.01.

Nelson, James Lindemann. (1997). "Everything Includes Itself in Power", in *Reading Engelhardt: Essays on the Thought of H. Tristram Engelhardt, Jr.*, Brendan P. Minogue, Gabriel Palmer-Fernandez, and James E. Reagan, eds. Boston, MA: Kluwer Academic Publishers, pp. 15–29.

Orlans, F. Barbara, Tom L. Beauchamp, Rebecca Dresser, David B. Morton, and John P. Gluck. (1998). *The Human Use of Animals*. New York: Oxford University Press.

Parfit, Derek. (1984). *Reasons and Persons*. Oxford: Clarendon Press.

Pellegrino, Edmund D. (1992). "Doctors Must Not Kill." *The Journal of Clinical Ethics* 3(2): 95–102.

Pellegrino, Edmund D. (1996). "Distortion of the Healing Relationship", in *Ethical Issues in Death and Dying*, 2nd ed. Tom L. Beauchamp and Robert Veatch, eds. Prentice Hall: Upper Saddle River, NJ, pp. 181–195.

Pellegrino, Edmund D. (2001). "The Internal Morality of Clinical Medicine: A Paradigm for the Ethics of the Helping and Healing Professions." *Journal of Medicine and Philosophy* 26(6): 559–579.

Perry, Joshua E., Larry R. Churchill, and Howard S. Kirshner (2005). "The Terri Schiavo Case: Legal, Ethical, and Medical Perspectives." *Annals of Internal Medicine* 143(10):744–748.

Quill, Timothy E. (1996). "Death and Dignity: A Case of Individualized Decision Making", in *Ethical Issues in Death and Dying*, 2nd ed. Tom L. Beauchamp and Robert M. Veatch, eds. Upper Saddle River, NJ: Prentice Hall, pp. 156–160.

Rachels, James. (1986). "Why Animals have a Right to Liberty", in Regan, Tom and Peter Singer, eds. *Animal Rights and Human Obligations*, 2nd ed. Englewood Cliffs, NJ: Prentice Hall, Inc., pp. 122–131.

Rawls, John. (1971). *A Theory of Justice*. Cambridge, MA: Harvard University Press.

Rawls, John. (1975). "The Independence of Moral Theory." *Proceedings and Addresses of the American Philosophical Association*, 48: 5–22.

Richardson, Henry. (1990). "Specifying Norms as a Way to Resolve Concrete Ethical Problems." *Philosophy and Public Affairs* 19: 279–310.

Richardson, Henry. (2000). "Specifying, Balancing, and Interpreting Ethical Principles." *Journal of Medicine and Philosophy* 25: 285–307.

Sandel, Michael. (1998). *Liberalism and the Limits of Justice*, 2nd ed. Cambridge: Cambridge University Press.

Sidgwick, Henry. (1981). *The Methods of Ethics*, 7th ed. (1907). Indianapolis: Hackett Publishing Co.

Strong, Carson. (1988). "Justification in Ethics", in *Moral Theory and Moral Judgments in Medical Ethics*, Baruch Brody, ed. Dordrecht: Kluwer Press, pp. 193–211.

Strong, Carson. (1999). "Critiques of Casuistry and Why They Are Mistaken." *Theoretical Medicine and Bioethics* 20: 395–411.

Strong, Carson. (2000) "Specified Principlism: What is it, and Does it Really Resolve Cases Better than Casuistry?" *Journal of Medicine and Philosophy* 25(3): 323–341.

Sunstein, Cass R. (1993). "On Analogical Reasoning." *Harvard Law Review* 106: 741–791.

Supreme Court of New Jersey. (1977). "In the Matter of Karen Quinlan, An Alleged Incompetent", in *Ethical Issues in Death and Dying,* 1st ed., Robert F. Weir, ed. New York: Columbia University Press, pp. 274–277.

Tallmon, James M. (1994). "How Jonsen Really Views Casuistry: A Note on the Abuse of Father Wildes." *Journal of Medicine and Philosophy* 19: 103–113.

Thomson, Judith Jarvis. (1997). "A Defense of Abortion", in *The Problem of Abortion,* 3rd ed., Susan Dwyer and Joel Feinberg, eds. Belmont, CA: Wadsworth Press, pp. 75–87.

Tomlinson, Tom. (1994). "Casuistry in Medical Ethics: Rehabilitated, or Repeat Offender?" *Theoretical Medicine* 15: 5–20.

United States [Supreme Court] Reports 274. (1927). 1000–1002. Reprinted in abridged form in *Contemporary Issues in Bioethics*, 4th ed. (1994). Tom L. Beauchamp and LeRoy Walters, eds. Belmont, CA: Wadsworth Publishing Co, pp. 607–608.

Veatch, Robert M. (1995). "Resolving Conflicts Among Principles: Ranking, Balancing, and Specifying." *Kennedy Institute of Ethics Journal* 5(3): 199–218.

Veatch, Robert M. (2001). "The Impossibility of a Morality Internal to Medicine." *Journal of Medicine and Philosophy* 26: 621–642.

Walzer, Michael. (1994). *Thick and Thin: Moral Argument at Home and Abroad*. Notre Dame, IN: University of Notre Dame Press.

Wiggins, Osborne P. and Michael A. Schwartz. (2005). "Richard Zaner's Phenomenology of the Clinical Encounter." *Theoretical Medicine and Bioethics* 26(1): 73–87.

Wildes, Kevin Wm. (1993). "The Priesthood of Bioethics and the Return of Casuistry." *Journal of Medicine and Philosophy* 18: 33–49.

Wildes, Kevin Wm. (1994). "Respondeo: Method and Content in Casuistry." *Journal of Medicine and Philosophy* 19: 115–119.

Wildes, Kevin Wm. (1997). "Engelhardt's Communitarian Ethics: The Hidden Assumptions", in *Reading Engelhardt: Essays on the Thought of H. Tristram Engelhardt, Jr.*, Brendan P. Minogue, Gabriel Palmer-Fernandez, and James E. Reagan, eds. Boston, MA: Kluwer Academic Publishers, pp. 77–93.

Wildes, Kevin Wm., S.J. (2000). *Moral Acquaintances: Methodology in Bioethics*. Notre Dame, IN: University of Notre Dame Press.

Wolf, Clark. (1995). "Contemporary Property Rights, Lockean Provisos, and the Interests of Future Generations." *Ethics* 105: 791–818.

Wolfson, Jay. (2005). "Erring on the Side of Theresa Schiavo: Reflections of the Special Guardian ad Litem." *The Hastings Center Report* 35(3):16–19.

Zaner, Richard M. (1988). *Ethics and the Clinical Encounter*. Edgewater Cliffs, NJ: Prentice Hall.

# Index